A Student's Guide to Coding and Information Theory

This easy-to-read guide provides a concise introduction to the engineering background of modern communication systems, from mobile phones to data compression and storage. Background mathematics and specific engineering techniques are kept to a minimum, so that only a basic knowledge of high-school mathematics is needed to understand the material covered. The authors begin with many practical applications in coding, including the repetition code, the Hamming code, and the Huffman code. They then explain the corresponding information theory, from entropy and mutual information to channel capacity and the information transmission theorem. Finally, they provide insights into the connections between coding theory and other fields. Many worked examples are given throughout the book, using practical applications to illustrate theoretical definitions. Exercises are also included, enabling readers to double-check what they have learned and gain glimpses into more advanced topics, making this perfect for anyone who needs a quick introduction to the subject.

STEFAN M. MOSER is an Associate Professor in the Department of Electrical Engineering at the National Chiao Tung University (NCTU), Hsinchu, Taiwan, where he has worked since 2005. He has received many awards for his work and teaching, including the Best Paper Award for Young Scholars by the IEEE Communications Society and IT Society (Taipei/Tainan Chapters) in 2009, the NCTU Excellent Teaching Award, and the NCTU Outstanding Mentoring Award (both in 2007).

PO-NING CHEN is a Professor in the Department of Electrical Engineering at the National Chiao Tung University (NCTU). Amongst his awards, he has received the 2000 Young Scholar Paper Award from Academia Sinica. He was also selected as the Outstanding Tutor Teacher of NCTU in 2002, and he received the Distinguished Teaching Award from the College of Electrical and Computer Engineering in 2003.

A Student's Guide to Coding and Information Theory

STEFAN M. MOSER

PO-NING CHEN

National Chiao Tung University (NCTU),
Hsinchu, Taiwan

CAMBRIDGE UNIVERSITY PRESS
Cambridge, New York, Melbourne, Madrid, Cape Town,
Singapore, São Paulo, Delhi, Mexico City

Cambridge University Press
The Edinburgh Building, Cambridge CB2 8RU, UK

Published in the United States of America by Cambridge University Press, New York

www.cambridge.org
Information on this title: www.cambridge.org/9781107015838

First published 2012

A catalogue record for this publication is available from the British Library

ISBN 978-1-107-01583-8 Hardback
ISBN 978-1-107-60196-3 Paperback

Additional resources for this publication at www.cambridge.org/moser

Contents

Contributors

Po-Ning Chen	(Chapter 7)
Francis Lu	(Chapter 3 and 8)
Stefan M. Moser	(Chapter 4 and 5)
Chung-Hsuan Wang	(Chapter 1 and 2)
Jwo-Yuh Wu	(Chapter 6)

Preface

Most of the books on coding and information theory are prepared for those who already have good background knowledge in probability and random processes. It is therefore hard to find a ready-to-use textbook in these two subjects suitable for engineering students at the freshmen level, or for non-engineering major students who are interested in knowing, at least conceptually, how information is encoded and decoded in practice and the theories behind it. Since communications has become a part of modern life, such knowledge is more and more of practical significance. For this reason, when our school requested us to offer a preliminary course in coding and information theory for students who do not have any engineering background, we saw this as an opportunity and initiated the plan to write a textbook.

In preparing this material, we hope that, in addition to the aforementioned purpose, the book can also serve as a beginner's guide that inspires and attracts students to enter this interesting area. The material covered in this book has been carefully selected to keep the amount of background mathematics and electrical engineering to a minimum. At most, simple calculus plus a little probability theory are used here, and anything beyond that is developed as needed. Its first version has been used as a textbook in the 2009 summer freshmen course *Conversion Between Information and Codes: A Historical View* at National Chiao Tung University, Taiwan. The course was attended by 47 students, including 12 from departments other than electrical engineering. Encouraged by the positive feedback from the students, the book went into a round of revision that took many of the students' comments into account. A preliminary version of this revision was again the basis of the corresponding 2010 summer freshmen course, which this time was attended by 51 students from ten different departments. Specific credit must be given to Professor Chung-Hsuan Wang, who volunteered to teach these 2009 and 2010 courses and whose input considerably improved the first version, to Ms. Hui-Ting

Chang (a graduate student in our institute), who has redrawn all the figures and brought them into shape, and to Pei-Yu Shih (a post-doc in our institute) and Ting-Yi Wu (a second-year Ph.D. student in our institute), who checked the readability and feasibility of all exercises. The authors also gratefully acknowledge the support from our department, which continues to promote this course.

Among the eight chapters in this book, Chapters 1 to 4 discuss coding techniques (including error-detecting and error-correcting codes), followed by a briefing in information theory in Chapters 5 and 6. By adopting this arrangement, students can build up some background knowledge on coding through concrete examples before plunging into information theory. Chapter 7 concludes the quest on information theory by introducing the Information Transmission Theorem. It attempts to explain the practical meaning of the so-called *Shannon limit* in communications, and reviews the historical breakthrough of turbo coding, which, after 50 years of research efforts, finally managed to approach this limit. The final chapter takes a few glances at unexpected relations between coding theory and other fields. This chapter is less important for an understanding of the basic principles, and is more an attempt to broaden the view on coding and information theory.

In summary, Chapter 1 gives an overview of this book, including the system model, some basic operations of information processing, and illustrations of how an information source is encoded.

Chapter 2 looks at ways of encoding source symbols such that any errors, up to a given level, can be detected at the receiver end. Basics of modular arithmetic that will be used in the analysis of the error-detecting capability are also included and discussed.

Chapter 3 introduces the fundamental concepts of error-correcting codes using the three-times repetition code and the Hamming code as starting examples. The error-detecting and -correcting capabilities of general linear block codes are also discussed.

Chapter 4 looks at data compression. It shows how source codes represent the output of an information source efficiently. The chapter uses Professor James L. Massey's beautifully simple and elegant approach based on trees. By this means it is possible to prove all main results in an intuitive fashion that relies on graphical explanations and requires no abstract math.

Chapter 5 presents a basic introduction to information theory and its main quantity *entropy*, and then demonstrates its relation to the source coding of Chapter 4. Since the basic definition of entropy and some of its properties are rather dry mathematical derivations, some time is spent on motivating the definitions. The proofs of the fundamental source coding results are then again based on trees and are therefore scarcely abstract in spite of their theoretical importance.

Chapter 6 addresses how to convey information reliably over a noisy communication channel. The *mutual information* between channel input and output is defined and then used to quantify the maximal amount of information that can get through a channel (the so-called *channel capacity*). The issue of how to achieve channel capacity via proper selection of the input is also discussed.

Chapter 7 begins with the introduction of the Information Transmission Theorem over communication channels corrupted by additive white Gaussian noise. The optimal error rate that has been proven to be attainable by Claude E. Shannon (baptized the *Shannon limit*) is then addressed, particularly for the situation when the amount of transmitted information is above the channel capacity. The chapter ends with a simple illustration of turbo coding, which is considered the first practical design approaching the Shannon limit.

Chapter 8 describes two particularly interesting connections between coding theory and seemingly unrelated fields: firstly the relation of the Hamming code to projective geometry is discussed, and secondly an application of codes to game theory is given.

The title, *A Student's Guide to Coding and Information Theory*, expresses our hope that this book is suitable as a beginner's guide, giving an overview to anyone who wishes to enter this area. In order not to scare the students (especially those without an engineering background), no problems are given at the end of each chapter as usual textbooks do. Instead, the problems are incorporated into the main text in the form of Exercises. The readers are encouraged to work them out. They are very helpful in understanding the concepts and are motivating examples for the theories covered in this book at a more advanced level.

The book will undergo further revisions as long as the course continues to be delivered. If a reader would like to provide comments or correct typos and errors, please email any of the authors. We will appreciate it very much!

1
Introduction

Systems dedicated to the communication or storage of information are commonplace in everyday life. Generally speaking, a communication system is a system which sends information from one place to another. Examples include telephone networks, computer networks, audio/video broadcasting, etc. Storage systems, e.g. magnetic and optical disk drives, are systems for storage and later retrieval of information. In a sense, such systems may be regarded as communication systems which transmit information from now (the present) to then (the future). Whenever or wherever problems of information processing arise, there is a need to know how to compress the textual material and how to protect it against possible corruption. This book is to cover the fundamentals of *information theory* and *coding theory*, to solve the above main problems, and to give related examples in practice. The amount of background mathematics and electrical engineering is kept to a minimum. At most, simple results of calculus and probability theory are used here, and anything beyond that is developed as needed.

1.1 Information theory versus coding theory

Information theory is a branch of probability theory with extensive applications to communication systems. Like several other branches of mathematics, information theory has a physical origin. It was initiated by communication scientists who were studying the statistical structure of electrical communication equipment and was principally founded by Claude E. Shannon through the landmark contribution [Sha48] on the mathematical theory of communications. In this paper, Shannon developed the fundamental limits on data compression and reliable transmission over noisy channels. Since its inception, information theory has attracted a tremendous amount of research effort and provided lots of inspiring insights into many research fields, not only communication and signal processing in electrical engineering, but also statistics, physics, computer science, economics, biology, etc.

Coding theory is mainly concerned with explicit methods for efficient and reliable data transmission or storage, which can be roughly divided into data compression and error-control techniques. Of the two, the former attempts to compress the data from a source in order to transmit or store them more efficiently. This practice is found every day on the Internet where data are usually transformed into the ZIP format to make files smaller and reduce the network load.

The latter adds extra data bits to make the transmission of data more robust to channel disturbances. Although people may not be aware of its existence in many applications, its impact has been crucial to the development of the Internet, the popularity of compact discs (CD), the feasibility of mobile phones, the success of the deep space missions, etc.

Logically speaking, coding theory leads to information theory, and information theory provides the performance limits on what can be done by suitable encoding of the information. Thus the two theories are intimately related, although in the past they have been developed to a great extent quite separately. One of the main purposes of this book is to show their mutual relationships.

1.2 Model and basic operations of information processing systems

Communication and storage systems can be regarded as examples of information processing systems and may be represented abstractly by the block diagram in Figure 1.1. In all cases, there is a *source* from which the information originates. The information source may be many things; for example, a book, music, or video are all information sources in daily life.

Figure 1.1 Basic information processing system.

The source output is processed by an *encoder* to facilitate the transmission (or storage) of the information. In communication systems, this function is often called a *transmitter*, while in storage systems we usually speak of a *recorder*. In general, three basic operations can be executed in the encoder: source coding, channel coding, and modulation. For source coding, the encoder maps the source output into digital format. The mapping is one-to-one, and the objective is to eliminate or reduce the redundancy, i.e. that part of the data which can be removed from the source output without harm to the information to be transmitted. So, source coding provides an efficient representation of the source output. For channel coding, the encoder introduces extra redundant data in a prescribed fashion so as to combat the noisy environment in which the information must be transmitted or stored. Discrete symbols may not be suitable for transmission over a physical channel or recording on a digital storage medium. Therefore, we need proper modulation to convert the data after source and channel coding to waveforms that are suitable for transmission or recording.

The output of the encoder is then transmitted through some physical communication channel (in the case of a communication system) or stored in some physical storage medium (in the case of a storage system). As examples of the former we mention wireless radio transmission based on electromagnetic waves, telephone communication through copper cables, and wired high-speed transmission through fiber optic cables. As examples of the latter we indicate magnetic storage media, such as those used by a magnetic tape, a hard-drive, or a floppy disk drive, and optical storage disks, such as a CD-ROM[1] or a DVD.[2] Each of these examples is subject to various types of noise disturbances. On a telephone line, the disturbance may come from thermal noise, switching noise, or crosstalk from other lines. On magnetic disks, surface defects and dust particles are regarded as noise disturbances. Regardless of the explicit form of the medium, we shall refer to it as the *channel*.

Information conveyed through (or stored in) the channel must be recovered at the destination and processed to restore its original form. This is the task of the *decoder*. In the case of a communication system, this device is often referred to as the *receiver*. In the case of a storage system, this block is often called the *playback system*. The signal processing performed by the decoder can be viewed as the inverse of the function performed by the encoder. The output of the decoder is then presented to the final user, which we call the *information sink*.

The physical channel usually produces a received signal which differs from the original input signal. This is because of signal distortion and noise introduced by the channel. Consequently, the decoder can only produce an estimate

[1] CD-ROM stands for *compact disc read-only memory*.
[2] DVD stands for *digital video disc* or *digital versatile disc*.

of the original information message. All well-designed systems aim at reproducing as reliably as possible while sending as much information as possible per unit time (for communication systems) or per unit storage (for storage systems).

1.3 Information source

Nature usually supplies information in continuous forms like, e.g., a beautiful mountain scene or the lovely chirping of birds. However, digital signals in which both amplitude and time take on discrete values are preferred in modern communication systems. Part of the reason for this use of digital signals is that they can be transmitted more reliably than analog signals. When the inevitable corruption of the transmission system begins to degrade the signal, the digital pulses can be detected, reshaped, and amplified to standard form before relaying them to their final destination. Figure 1.2 illustrates an ideal binary digital pulse propagating along a transmission line, where the pulse shape is degraded as a function of line length. At a propagation distance where the transmitted pulse can still be reliably identified (before it is degraded to an ambiguous state), the pulse is amplified by a digital amplifier that recovers its original ideal shape. The pulse is thus regenerated. On the other hand, analog signals cannot be so reshaped since they take an infinite variety of shapes. Hence the farther the signal is sent and the more it is processed, the more degradation it suffers from small errors.

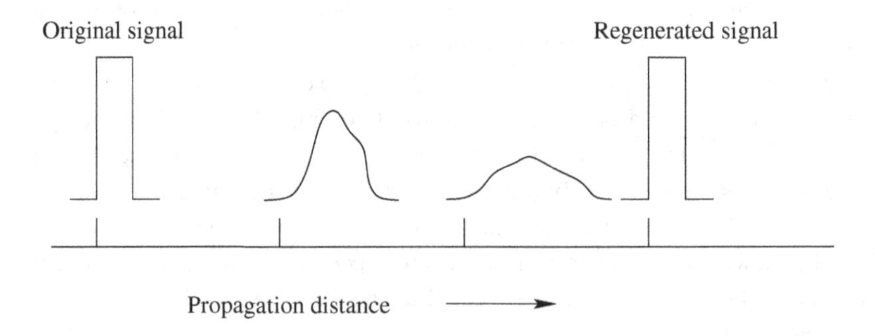

Figure 1.2 Pulse degradation and regeneration.

Modern practice for transforming analog signals into digital form is to sample the continuous signal at equally spaced intervals of time, and then to quantize the observed value, i.e. each sample value is approximated by the nearest level in a finite set of discrete levels. By mapping each quantized sample to a codeword consisting of a prescribed number of code elements, the information is then sent as a stream of digits. The conversion process is illustrated in Figure 1.3. Figure 1.3(a) shows a segment of an analog waveform. Figure 1.3(b) shows the corresponding digital waveform based on the binary code in Table 1.1. In this example, symbols 0 and 1 of the binary code are represented by zero and one volt, respectively. Each sampled value is quantized into four binary digits (bits) with the last bit called *sign bit* indicating whether the sample value is positive or negative. The remaining three bits are chosen to represent the absolute value of a sample in accordance with Table 1.1.

Table 1.1 *Binary representation of quantized levels*

Index of quantization level	Binary representation	Index expressed as sum of powers of 2
0	000	
1	001	2^0
2	010	2^1
3	011	$2^1 + 2^0$
4	100	2^2
5	101	$2^2 \quad + 2^0$
6	110	$2^2 + 2^1$
7	111	$2^2 + 2^1 + 2^0$

As a result of the sampling and quantizing operations, errors are introduced into the digital signal. These errors are nonreversible in that it is not possible to produce an exact replica of the original analog signal from its digital representation. However, the errors are under the designer's control. Indeed, by proper selection of the sampling rate and number of the quantization levels, the errors due to the sampling and quantizing can be made so small that the difference between the analog signal and its digital reconstruction is not discernible by a human observer.

1.4 Encoding a source alphabet

Based on the discussion in Section 1.3, we can assume without loss of generality that an information source generates a finite (but possibly large) number of messages. This is undoubtedly true for a digital source. As for an analog

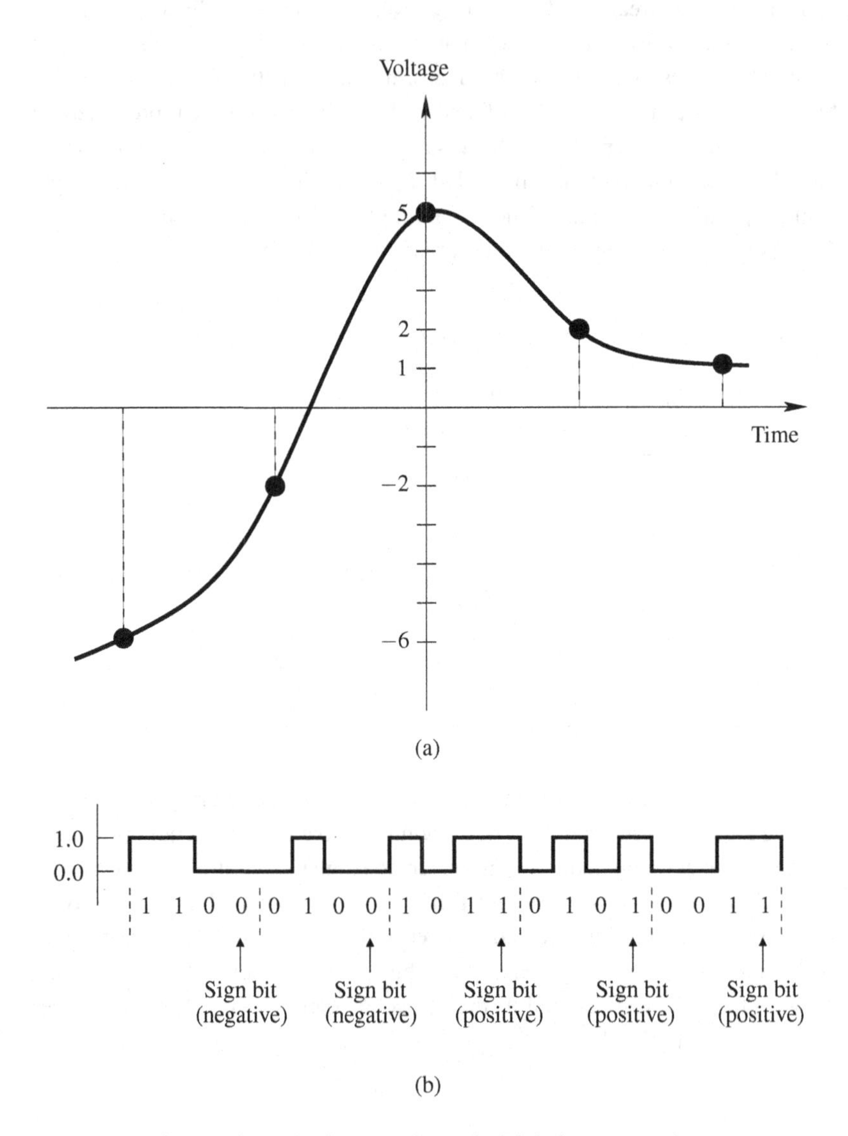

Figure 1.3 (a) Analog waveform. (b) Digital representation.

source, the analog-to-digital conversion process mentioned above also makes the assumption feasible. However, even though specific messages are actually sent, the system designer has no idea in advance which message will be chosen for transmission. We thus need to think of the source as a random (or stochastic) source of information, and ask how we may encode, transmit, and recover the original information.

An information source's output alphabet is defined as the collection of all possible messages. Denote by $\mathcal{U}$ a source alphabet which consists of r messages, say $u_1, u_2, \ldots, u_r$, with probabilities $p_1, p_2, \ldots, p_r$ satisfying

$$p_i \geq 0, \ \forall i, \ \text{and} \ \sum_{i=1}^{r} p_i = 1. \tag{1.1}$$

Here the notation $\forall$ means "for all" or "for every." We can always represent each message by a sequence of bits, which provides for easier processing by computer systems. For instance, if we toss a fair dice to see which number faces up, only six possible outputs are available with $\mathcal{U} = \{1,2,3,4,5,6\}$ and $p_i = 1/6, \forall\, 1 \leq i \leq 6$. The following shows a straightforward binary description of these messages:

$$1 \leftrightarrow 001,\ 2 \leftrightarrow 010,\ 3 \leftrightarrow 011,\ 4 \leftrightarrow 100,\ 5 \leftrightarrow 101,\ 6 \leftrightarrow 110, \tag{1.2}$$

where each decimal number is encoded as its binary expression. Obviously, there exist many other ways of encoding. For example, consider the two mappings listed below:

$$1 \leftrightarrow 00,\ 2 \leftrightarrow 01,\ 3 \leftrightarrow 100,\ 4 \leftrightarrow 101,\ 5 \leftrightarrow 110,\ 6 \leftrightarrow 111 \tag{1.3}$$

and

$$1 \leftrightarrow 1100,\ 2 \leftrightarrow 1010,\ 3 \leftrightarrow 0110,\ 4 \leftrightarrow 1001,\ 5 \leftrightarrow 0101,\ 6 \leftrightarrow 0011. \tag{1.4}$$

Note that all the messages are one-to-one mapped to the binary sequences, no matter which of the above encoding methods is employed. The original message can always be recovered from the binary sequence.

Given an encoding method, let l_i denote the length of the output sequence, called the *codeword*, corresponding to u_i, $\forall\, 1 \leq i \leq r$. From the viewpoint of source coding for data compression, an optimal encoding should minimize the average length of codewords defined by

$$\mathsf{L}_{\text{av}} \triangleq \sum_{i=1}^{r} p_i l_i. \tag{1.5}$$

By (1.5), the average lengths of codewords in (1.2), (1.3), and (1.4) are, respectively,

$$\mathrm{L}_{\mathrm{av}}^{(1.2)} = \frac{1}{6}3 + \frac{1}{6}3 + \frac{1}{6}3 + \frac{1}{6}3 + \frac{1}{6}3 + \frac{1}{6}3 = 3, \tag{1.6}$$

$$\mathrm{L}_{\mathrm{av}}^{(1.3)} = \frac{1}{6}2 + \frac{1}{6}2 + \frac{1}{6}3 + \frac{1}{6}3 + \frac{1}{6}3 + \frac{1}{6}3 = \frac{8}{3} \simeq 2.667, \tag{1.7}$$

$$\mathrm{L}_{\mathrm{av}}^{(1.4)} = \frac{1}{6}4 + \frac{1}{6}4 + \frac{1}{6}4 + \frac{1}{6}4 + \frac{1}{6}4 + \frac{1}{6}4 = 4. \tag{1.8}$$

The encoding method in (1.3) thus provides a more efficient way for the representation of these source messages.

As for channel coding, a good encoding method should be able to protect the source messages against the inevitable noise corruption. Suppose 3 is to be transmitted and an error occurs in the least significant bit (LSB), namely the first bit counted from the right-hand side of the associated codeword. In the case of code (1.2) we now receive 010 instead of 011, and in the case of code (1.3) we receive 101 instead of 100. In both cases, the decoder will retrieve a wrong message (2 and 4, respectively). However, 0111 will be received if 3 is encoded by (1.4). Since 0111 is different from all the codewords in (1.4), we can be aware of the occurrence of an error, i.e. the error is detected, and possible retransmission of the message can be requested. Not just the error in the LSB, but any single error can be detected by this encoding method. The code (1.4) is therefore a better choice from the viewpoint of channel coding.

Typically, for channel coding, the encoding of the message to be transmitted over the channel adds redundancy to combat channel noise. On the other hand, the source encoding usually removes redundancy contained in the message to be compressed. A more detailed discussion on channel and source coding will be shown in Chapters 2 and 3 and in Chapters 4 and 5, respectively.

1.5 Octal and hexadecimal codes

Although the messages of an information source are usually encoded as binary sequences, the binary code is sometimes inconvenient for humans to use. People usually prefer to make a single discrimination among many things. Evidence for this is the size of the common alphabets. For example, the English alphabet has 26 letters, the Chinese "alphabet" (bopomofo) has 37 letters, the Phoenician alphabet has 22 letters, the Greek alphabet has 24 letters, the Russian alphabet 33, the Cyrillic alphabet has 44 letters, etc. Thus, for human use, it is often convenient to group the bits into groups of three at a time and call them the octal code (base 8). This code is given in Table 1.2.

Table 1.2 *Octal code*

Binary	Octal
000	0
001	1
010	2
011	3
100	4
101	5
110	6
111	7

When using the octal representation, numbers are often enclosed in parentheses with a following subscript 8. For example, the decimal number 81 is written in octal as $(121)_8$ since $81 = \text{“1”} \times 8^2 + \text{“2”} \times 8^1 + \text{“1”} \times 8^0$. The translation from octal to binary is so immediate that there is little trouble in going either way.

The binary digits are sometimes grouped in fours to make the hexadecimal code (Table 1.3). For instance, to translate the binary sequence 101011000111 to the octal form, we first partition these bits into groups of three:

$$\underbrace{101}\underbrace{011}\underbrace{000}\underbrace{111}. \tag{1.9}$$

Each group of bits is then mapped to an octal number by Table 1.2, hence resulting in the octal representation $(5307)_8$. If we partition the bits into groups of four, i.e.

$$\underbrace{1010}\underbrace{1100}\underbrace{0111}, \tag{1.10}$$

we can get the hexadecimal representation $(\mathrm{AC7})_{16}$ by Table 1.3. Since computers usually work in *bytes*, which are 8 bits each, the hexadecimal code fits into the machine architecture better than the octal code. However, the octal code seems to fit better into the human's psychology. Thus, in practice, neither code has a clear victory over the other.

1.6 Outline of the book

After the introduction of the above main topics, we now have a basis for discussing the material the book is to cover.

Table 1.3 *Hexadecimal code*

Binary	Hexadecimal
0000	0
0001	1
0010	2
0011	3
0100	4
0101	5
0110	6
0111	7
1000	8
1001	9
1010	A
1011	B
1100	C
1101	D
1110	E
1111	F

In general, the error-detecting capability will be accomplished by adding some digits to the message, thus making the message slightly longer. The main problem is to achieve a required protection against the inevitable channel errors without too much cost in adding extra digits. Chapter 2 will look at ways of encoding source symbols so that any errors, up to a given level, may be detected at the terminal end. For a detected error, we might call for a repeat transmission of the message, hoping to get it right the next time.

In contrast to error-detecting codes, error-correcting codes are able to correct some detected errors directly without having to retransmit the message a second time. In Chapter 3, we will discuss two kinds of error-correcting codes, the *repetition code* and the *Hamming code*, as well as their encoding and decoding methods.

In Chapter 4, we consider ways of representing information in an efficient way. The typical example will be an information source that can take on r different possible values. We will represent each of these r values by a string of 0s and 1s with varying length. The question is how to design these strings such that the average length is minimized, but such that we are still able to recover the original data from it. So, in contrast to Chapters 2 and 3, here we try to shorten the codewords.

While in Chapters 2 to 4 we are concerned with coding theory, Chapter 5 introduces information theory. We define some way of measuring "information" and then apply it to the codes introduced in Chapter 4. By doing so we can not only compare different codes but also derive some fundamental limits of what is possible and what not. So Chapter 5 provides the information theory related to the coding theory introduced in Chapter 4.

In Chapter 6, we continue on the path of information theory and develop the relation to the coding theory of Chapters 2 and 3. Prior to the mid 1940s people believed that transmitted data subject to noise corruption can never be perfectly recovered unless the transmission rate approaches zero. Shannon's landmark work in 1948 [Sha48] disproved this thinking and established a fundamental result for modern communication: as long as the transmission rate is below a certain threshold (the so-called *channel capacity*), errorless data transmission can be realized by some properly designed coding scheme. Chapter 6 will highlight the essentials regarding the channel capacity. We shall first introduce a communication channel model from the general probabilistic setting. Based on the results of Chapter 5, we then go on to specify the mutual information, which provides a natural way of characterizing the channel capacity.

In Chapter 7, we build further on the ideas introduced in Chapters 2, 3, and 6. We will cover the basic concept of the theory of reliable transmission of information bearing signals over a noisy communication channel. In particular, we will discuss the *additive white Gaussian noise (AWGN) channel* and introduce the famous *turbo code* that is the first code that can approach the Shannon limit of the AWGN channel up to less than 1 dB at a bit error rate (BER) of 10^{-5}.

Finally, in Chapter 8, we try to broaden the view by showing two relations of coding theory to quite unexpected fields. Firstly we explain a connection of projective geometry to the Hamming code of Chapter 3. Secondly we show how codes (in particular the three-times repetition code and the Hamming code) can be applied to game theory.

References

[Sha48] Claude E. Shannon, "A mathematical theory of communication," *Bell System Technical Journal*, vol. 27, pp. 379–423 and 623–656, July and October 1948. Available: http://moser.cm.nctu.edu.tw/nctu/doc/shannon1948.pdf

2

Error-detecting codes

When a message is transmitted, the inevitable noise disturbance usually degrades the quality of communication. Whenever repetition is possible, it is sufficient to detect the occurrence of an error. When an error is detected, we simply repeat the message, and it may be correct the second time or even possibly the third time.

It is not possible to detect an error if every possible symbol, or set of symbols, that can be received is a legitimate message. It is only possible to catch errors if there are some restrictions on what a proper message is. The problem is to keep these restrictions on the possible messages down to ones that are simple. In practice, "simple" means "easily computable." In this chapter, we will mainly investigate the problem of designing codes such that any single error can be detected at the receiver. In Chapter 3, we will then consider *correcting* the errors that occur during the transmission.

2.1 Review of modular arithmetic

We first give a quick review of the basic arithmetic which is extensively used in the following sections. For binary digits, which take values of only 0 and 1, the rules for addition and multiplication are defined by

$$\begin{aligned} 0+0&=0 \\ 0+1&=1 \\ 1+0&=1 \\ 1+1&=0 \end{aligned} \qquad \text{and} \qquad \begin{aligned} 0\times 0&=0 \\ 0\times 1&=0 \\ 1\times 0&=0 \\ 1\times 1&=1, \end{aligned} \tag{2.1}$$

respectively. For example, by (2.1), we have

$$1+1\times 0+0+1\times 1=1+0+0+1=0. \tag{2.2}$$

If we choose to work in the decimal arithmetic, the binary arithmetic in (2.1) can be obtained by dividing the result in decimal by 2 and taking the remainder. For example, (2.2) yields

$$1+0+0+1=2\equiv 0 \bmod 2. \tag{2.3}$$

Occasionally, we may work modulo some number other than 2 for the case of a nonbinary source. Given a positive integer m, for the addition and multiplication mod m ("mod" is an abbreviation for "modulo"), we merely divide the result in decimal by m and take the nonnegative remainder. For instance, consider an information source with five distinct outputs 0, 1, 2, 3, 4. It follows that

$$2+4=1\times 5+\text{“1”} \quad\Longleftrightarrow\quad 2+4\equiv 1 \bmod 5, \tag{2.4}$$

$$3\times 4=2\times 5+\text{“2”} \quad\Longleftrightarrow\quad 3\times 4\equiv 2 \bmod 5. \tag{2.5}$$

Other cases for the modulo 5 addition and multiplication can be referred to in Table 2.1.

Table 2.1 *Addition and multiplication modulo 5*

+ mod 5	0	1	2	3	4
0	0	1	2	3	4
1	1	2	3	4	0
2	2	3	4	0	1
3	3	4	0	1	2
4	4	0	1	2	3

× mod 5	0	1	2	3	4
0	0	0	0	0	0
1	0	1	2	3	4
2	0	2	4	1	3
3	0	3	1	4	2
4	0	4	3	2	1

For multiplication mod m, we have to be more careful if m is not a prime. Suppose that we have the numbers a and b congruent to a' and b' modulo the modulus m. This means that

$$a\equiv a' \bmod m \quad \text{and} \quad b\equiv b' \bmod m \tag{2.6}$$

or

$$a=a'+k_1 m \quad \text{and} \quad b=b'+k_2 m \tag{2.7}$$

for some integers k_1 and k_2. For the product ab, we have

$$ab=a'b'+a'k_1 m+b'k_2 m+k_1 k_2 m^2 \tag{2.8}$$

and hence

$$ab\equiv a'b' \bmod m. \tag{2.9}$$

Now consider the particular case

$$a = 15, \quad b = 12, \quad m = 10. \tag{2.10}$$

We have $a' = 5$ and $b' = 2$ by (2.7) and $ab \equiv a'b' \equiv 0 \bmod 10$ by (2.9). But neither a nor b is zero! Only for a *prime* modulus do we have the important property that if a product is zero, then at least one factor must be zero.

Exercise 2.1 *In order to become more familiar with the modular operation check out the following problems:*

$$3 \times 6 + 7 \equiv ? \bmod 11 \tag{2.11}$$

and

$$5 - 4 \times 2 \equiv ? \bmod 7. \tag{2.12}$$

◇

More on modular arithmetic can be found in Section 3.1.

2.2 Independent errors – white noise

To simplify the analysis of noise behavior, we assume that errors in a message satisfy the following constraints:

(1) the probability of an error in any binary position is assumed to be a fixed number p, and
(2) errors in different positions are assumed to be independent.[1]

Such noise is called "white noise" in analogy with white light, which is supposed to contain uniformly all the frequencies detected by the human eye. However, in practice, there are often reasons for errors to be more common in some positions in the message than in others, and it is often true that errors tend to occur in bursts and not to be independent.We assume white noise in the very beginning because this is the simplest case, and it is better to start from the simplest case and move on to more complex situations after we have built up a solid knowledge on the simple case.

Consider a message consisting of n digits for transmission. For white noise, the probability of no error in any position is given by

$$(1-p)^n. \tag{2.13}$$

[1] Given events $\mathcal{A}_\ell$, they are said to be independent if $\Pr\left(\bigcap_{\ell=1}^n \mathcal{A}_\ell\right) = \prod_{\ell=1}^n \Pr(\mathcal{A}_\ell)$. Here "$\bigcap_\ell$" denotes set-intersection, i.e. $\bigcap_\ell \mathcal{A}_\ell$ is the set of elements that are members of *all* sets $\mathcal{A}_\ell$. Hence, $\Pr(\bigcap_{\ell=1}^n \mathcal{A}_\ell)$ is the event that *all* events $\mathcal{A}_\ell$ occur at the same time. The notation $\prod_\ell$ is a shorthand for multiplication: $\prod_{\ell=1}^n a_\ell \triangleq a_1 \cdot a_2 \cdots a_n$.

The probability of a single error in the message is given by

$$np(1-p)^{n-1}. \tag{2.14}$$

The probability of ℓ errors is given by the $(\ell+1)$th term in the binomial expansion:

$$\begin{aligned}
1 &= ((1-p)+p)^n \qquad (2.15)\\
&= \binom{n}{0}(1-p)^n + \binom{n}{1}p(1-p)^{n-1} + \binom{n}{2}p^2(1-p)^{n-2}\\
&\quad + \cdots + \binom{n}{n}p^n \qquad (2.16)\\
&= (1-p)^n + np(1-p)^{n-1} + \frac{n(n-1)}{2}p^2(1-p)^{n-2} + \cdots + p^n. \qquad (2.17)
\end{aligned}$$

For example, the probability of exactly two errors is given by

$$\frac{n(n-1)}{2}p^2(1-p)^{n-2}. \tag{2.18}$$

We can obtain the probability of an even number of errors $(0,2,4,\ldots)$ by adding the following two binomial expansions and dividing by 2:

$$1 = ((1-p)+p)^n = \sum_{\ell=0}^{n}\binom{n}{\ell}p^\ell(1-p)^{n-\ell}, \tag{2.19}$$

$$(1-2p)^n = ((1-p)-p)^n = \sum_{\ell=0}^{n}(-1)^\ell\binom{n}{\ell}p^\ell(1-p)^{n-\ell}. \tag{2.20}$$

Denote by $\lfloor\xi\rfloor$ the greatest integer not larger than ξ. We have[2]

$$\begin{aligned}
\Pr(\text{An even number of errors}) &= \sum_{\ell=0}^{\lfloor n/2\rfloor}\binom{n}{2\ell}p^{2\ell}(1-p)^{n-2\ell} \qquad (2.21)\\
&= \frac{1+(1-2p)^n}{2}. \qquad (2.22)
\end{aligned}$$

The probability of an odd number of errors is 1 minus this number.

Exercise 2.2 *Actually, this is a good chance to practice your basic skills on the method of induction: can you show that*

$$\sum_{\ell=0}^{\lfloor n/2\rfloor}\binom{n}{2\ell}p^{2\ell}(1-p)^{n-2\ell} = \frac{1+(1-2p)^n}{2} \tag{2.23}$$

and

$$\sum_{\ell=0}^{\lfloor (n-1)/2\rfloor}\binom{n}{2\ell+1}p^{2\ell+1}(1-p)^{n-2\ell-1} = \frac{1-(1-2p)^n}{2} \tag{2.24}$$

[2] Note that zero errors also counts as an even number of errors here.

by induction on n?

Hint: Note that

$$\binom{n+1}{k} = \binom{n}{k} + \binom{n}{k-1} \tag{2.25}$$

for $n,k \geq 1$. ◊

2.3 Single parity-check code

The simplest way of encoding a binary message to make it error-detectable is to count the number of 1s in the message, and then append a final binary digit chosen so that the entire message has an even number of 1s in it. The entire message is therefore of *even parity*. Thus to $(n-1)$ message positions we append an nth parity-check position. Denote by x_ℓ the original bit in the ℓth message position, $\forall 1 \leq \ell \leq n-1$, and let x_n be the parity-check bit. The constraint of even parity implies that

$$x_n = \sum_{\ell=1}^{n-1} x_\ell \tag{2.26}$$

by (2.1). Note that here (and for the remainder of this book) we omit "mod 2" and implicitly assume it everywhere. Let y_ℓ be the channel output corresponding to x_ℓ, $\forall 1 \leq \ell \leq n$. At the receiver, we firstly count the number of 1s in the received sequence $\mathbf{y}$. If the even-parity constraint is violated for the received vector, i.e.

$$\sum_{\ell=1}^{n} y_\ell \neq 0, \tag{2.27}$$

this indicates that at least one error has occurred.

For example, given a message $(x_1, x_2, x_3, x_4) = (0\,1\,1\,1)$, the parity-check bit is obtained by

$$x_5 = 0 + 1 + 1 + 1 = 1, \tag{2.28}$$

and hence the resulting even-parity codeword $(x_1, x_2, x_3, x_4, x_5)$ is $(0\,1\,1\,1\,1)$. Suppose the codeword is transmitted, but a vector $\mathbf{y} = (0\,0\,1\,1\,1)$ is received. In this case, an error in the second position is met. We have

$$y_1 + y_2 + y_3 + y_4 + y_5 = 0 + 0 + 1 + 1 + 1 = 1\,(\neq 0); \tag{2.29}$$

thereby the error is detected. However, if another vector of $(0\,0\,1\,1\,0)$ is received, where two errors (in the second and the last position) have occurred, no error will be detected since

$$y_1+y_2+y_3+y_4+y_5=0+0+1+1+0=0. \tag{2.30}$$

Evidently in this code any odd number of errors can be detected. But any even number of errors cannot be detected.

For channels with white noise, (2.22) gives the probability of any even number of errors in the message. Dropping the first term of (2.21), which corresponds to the probability of no error, we have the following probability of undetectable errors for the single parity-check code introduced here:

$$\Pr(\text{Undetectable errors}) = \sum_{\ell=1}^{\lfloor n/2 \rfloor} \binom{n}{2\ell} p^{2\ell}(1-p)^{n-2\ell} \tag{2.31}$$

$$= \frac{1+(1-2p)^n}{2} - (1-p)^n. \tag{2.32}$$

The probability of detectable errors, i.e. all the odd-number errors, is then obtained by

$$\Pr(\text{Detectable errors}) = 1 - \frac{1+(1-2p)^n}{2} = \frac{1-(1-2p)^n}{2}. \tag{2.33}$$

Obviously, we should have that

$$\Pr(\text{Detectable errors}) \gg \Pr(\text{Undetectable errors}). \tag{2.34}$$

For p very small, we have

$$\Pr(\text{Undetectable errors}) = \frac{1+(1-2p)^n}{2} - (1-p)^n \tag{2.35}$$

$$= \frac{1}{2} + \frac{1}{2}\left[\binom{n}{0} - \binom{n}{1}(2p) + \binom{n}{2}(2p)^2 - \cdots\right]$$
$$- \left[\binom{n}{0} - \binom{n}{1}p + \binom{n}{2}p^2 - \cdots\right] \tag{2.36}$$

$$= \frac{1}{2} + \frac{1}{2}\left[1 - 2np + \frac{n(n-1)}{2}4p^2 - \cdots\right]$$
$$- \left[1 - np + \frac{n(n-1)}{2}p^2 - \cdots\right] \tag{2.37}$$

$$\simeq \frac{n(n-1)}{2}p^2 \tag{2.38}$$

and

$$\Pr(\text{Detectable errors}) = \frac{1-(1-2p)^n}{2} \tag{2.39}$$

$$= \frac{1}{2} - \frac{1}{2}\left[\binom{n}{0} - \binom{n}{1}(2p) + \cdots\right] \tag{2.40}$$

$$= \frac{1}{2} - \frac{1}{2}[1 - 2np + \cdots] \tag{2.41}$$

$$\simeq np. \tag{2.42}$$

In the above approximations, we only retain the leading term that dominates the sum.

Hence, (2.34) requires

$$np \gg \frac{n(n-1)}{2}p^2, \tag{2.43}$$

and implies that the shorter the message, the better the detecting performance.

In practice, it is common to break up a long message in the binary alphabet into blocks of $(n-1)$ digits and to append one binary digit to each block. This produces the *redundancy* of

$$\frac{n}{n-1} = 1 + \frac{1}{n-1}, \tag{2.44}$$

where the redundancy is defined as the total number of binary digits divided by the minimum necessary. The *excess redundancy* is $1/(n-1)$. Clearly, for low redundancy we want to use long messages, but for high reliability short messages are better. Thus the choice of the length n for the blocks to be sent is a compromise between the two opposing forces.

2.4 The ASCII code

Here we introduce an example of a single parity-check code, called the *American Standard Code for Information Interchange (ASCII)*, which was the first code developed specifically for computer communications. Each character in ASCII is represented by seven data bits constituting a unique binary sequence. Thus a total of 128 $(= 2^7)$ different characters can be represented in ASCII. The characters are various commonly used letters, numbers, special control symbols, and punctuation symbols, e.g. \$, %, and @. Some of the special control symbols, e.g. ENQ (enquiry) and ETB (end of transmission block), are used for communication purposes. Other symbols, e.g. BS (back space) and CR (carriage return), are used to control the printing of characters on a page. A complete listing of ASCII characters is given in Table 2.2.

Since computers work in bytes which are blocks of 8 bits, a single ASCII symbol often uses 8 bits. The eighth bit is set so that the total number of 1s in the eight positions is an even number. For example, consider "K" in Table 2.2 encoded as $(113)_8$, which can be transformed into binary form as follows:

$$(113)_8 = 1001011 \tag{2.45}$$

Table 2.2 *Seven-bit ASCII code*

Octal code	Char.	Octal code	Char.	Octal code	Char.	Octal code	Char.
000	NUL	040	SP	100	@	140	‘
001	SOH	041	!	101	A	141	a
002	STX	042	”	102	B	142	b
003	ETX	043	#	103	C	143	c
004	EOT	044	$	104	D	144	d
005	ENQ	045	%	105	E	145	e
006	ACK	046	&	106	F	146	f
007	BEL	047	’	107	G	147	g
010	BS	050	(	110	H	150	h
011	HT	051	)	111	I	151	i
012	LF	052	*	112	J	152	j
013	VT	053	+	113	K	153	k
014	FF	054	,	114	L	154	l
015	CR	055	-	115	M	155	m
016	SO	056	.	116	N	156	n
017	SI	057	/	117	O	157	o
020	DLE	060	0	120	P	160	p
021	DC1	061	1	121	Q	161	q
022	DC2	062	2	122	R	162	r
023	DC3	063	3	123	S	163	s
024	DC4	064	4	124	T	164	t
025	NAK	065	5	125	U	165	u
026	SYN	066	6	126	V	166	v
027	ETB	067	7	127	W	167	w
030	CAN	070	8	130	X	170	x
031	EM	071	9	131	Y	171	y
032	SUB	072	:	132	Z	172	z
033	ESC	073	;	133	[	173	{
034	FS	074	<	134	\	174	|
035	GS	075	=	135	]	175	}
036	RS	076	>	136	^	176	~
037	US	077	?	137	_	177	DEL

(where we have dropped the first 2 bits of the first octal symbol). In this case, the parity-check bit is 0; "K" is thus encoded as 1001011$\underline{0}$ for even parity. You are encouraged to encode the remaining characters in Table 2.2.

By the constraint of even parity, any single error, a 0 changed into a 1 or a 1 changed into a 0, will be detected[3] since after the change there will be an odd number of 1s in the eight positions. Thus, we have an error-detecting code that helps to combat channel noise. Perhaps more importantly, the code makes it much easier to maintain the communication quality since the machine can detect the occurrence of errors by itself.

2.5 Simple burst error-detecting code

In some situations, errors occur in bursts rather than in isolated positions in the received message. For instance, lightning strikes, power-supply fluctuations, loose flakes on a magnetic surface are all typical causes of a burst of noise. Suppose that the maximum length of any error burst[4] that we are to detect is L. To protect data against the burst errors, we first divide the original message into a sequence of words consisting of L bits. Aided with a pre-selected error-detecting code, parity checks are then computed over the corresponding word positions, instead of the bit positions.

Based on the above scenario, if an error burst occurs within one word, in effect only a single word error is observed. If an error burst covers the end of one word and the beginning of another, still no two errors corresponding to the same position of words will be met, since we assumed that any burst length l satisfies $0 \leq l \leq \mathsf{L}$. Consider the following example for illustration.

Example 2.3 If the message is

Hello NCTU

and the maximum burst error length L is 8, we can use the 7-bit ASCII code in Table 2.2 and protect the message against burst errors as shown in Table 2.3. (Here no parity check is used within the ASCII symbols.) The encoded message is therefore

Hello NCTUn

[3] Actually, to be precise, every odd number of errors is detected.

[4] An error burst is said to have length L if errors are confined to L consecutive positions. By this definition, the *error patterns* 0111110, 0101010, and 0100010 are all classified as bursts of length 5. Note that a 0 in an error pattern denotes that no error has happened in that position, while a 1 denotes an error. See also (3.34) in Section 3.3.2.

Table 2.3 *Special type of parity check to protect against burst errors of maximum length* $\mathsf{L} = 8$

$\mathrm{H} = (110)_8 =$	01001000
$\mathrm{e} = (145)_8 =$	01100101
$\mathrm{l} = (154)_8 =$	01101100
$\mathrm{l} = (154)_8 =$	01101100
$\mathrm{o} = (157)_8 =$	01101111
$\mathrm{SP} = (040)_8 =$	00100000
$\mathrm{N} = (116)_8 =$	01001110
$\mathrm{C} = (103)_8 =$	01000011
$\mathrm{T} = (124)_8 =$	01010100
$\mathrm{U} = (125)_8 =$	01010101
Check sum =	01101110 = n

where n is the parity-check symbol.

Suppose a burst error of length 5, as shown in Table 2.4, is met during the transmission of the above message, where the bold-face positions are in error. In this case, the burst error is successfully detected since the check sum is not 00000000. However, if the burst error of length 16 shown in Table 2.5 occurs, the error will not be detected due to the all-zero check sum. ◊

Exercise 2.4 *Could you repeat the above process of encoding for the case of* $\mathsf{L} = 16$*? Also, show that the resulting code can detect all the bursts of length at most 16.* ◊

Exercise 2.5 *Can you show that the error might not be detected if there is more than one burst, even if each burst is of length no larger than* L*?* ◊

2.6 Alphabet plus number codes – weighted codes

The codes we have discussed so far were all designed with respect to a simple form of "white noise" that causes some bits to be flipped. This is very suitable for many types of machines. However, in some systems, where people are involved, other types of noise are more appropriate. The first common human error is to interchange adjacent digits of numbers; for example, 38 becomes 83. A second common error is to double the wrong one of a triple of digits, where two adjacent digits are the same; for example, 338 becomes 388. In addition, the confusion of O ("oh") and 0 ("zero") is also very common.

Table 2.4 *A burst error of length 5 has occurred during transmission and is detected because the check sum is not 0000000; bold-face positions denote positions in error*

H ⇒ K	0	1	0	0	1	0	**1**	**1**
e ⇒ ENQ	0	**0**	**0**	0	0	1	0	1
l	0	1	1	0	1	1	0	0
l	0	1	1	0	1	1	0	0
o	0	1	1	0	1	1	1	1
SP	0	0	1	0	0	0	0	0
N	0	1	0	0	1	1	1	0
C	0	1	0	0	0	0	1	1
T	0	1	0	1	0	1	0	0
U	0	1	0	1	0	1	0	1
n	0	1	1	0	1	1	1	0
Check sum =	0	1	1	0	0	0	1	1

Table 2.5 *A burst error of length 16 has occurred during transmission, but it is not detected; bold-face positions denote positions in error*

H ⇒ K	0	1	0	0	1	0	**1**	**1**
e ⇒ J	0	1	**0**	0	**1**	**0**	**1**	**0**
l ⇒ @	0	1	**0**	0	**0**	**0**	0	0
l	0	1	1	0	1	1	0	0
o	0	1	1	0	1	1	1	1
SP	0	0	1	0	0	0	0	0
N	0	1	0	0	1	1	1	0
C	0	1	0	0	0	0	1	1
T	0	1	0	1	0	1	0	0
U	0	1	0	1	0	1	0	1
n	0	1	1	0	1	1	1	0
Check sum =	0	0	0	0	0	0	0	0

Table 2.6 *Weighted sum: progressive digiting*

Message	Sum	Sum of sum
w	w	w
x	$w+x$	$2w+x$
y	$w+x+y$	$3w+2x+y$
z	$w+x+y+z$	$4w+3x+2y+z$

In English text-based systems, it is quite common to have a source alphabet consisting of the 26 letters, space, and the 10 decimal digits. Since the size of this source alphabet, 37 ($= 26+1+10$), is a prime number, we can use the following method to detect the presence of the above described typical errors. Firstly, each symbol in the source alphabet is mapped to a distinct number in $\{0,1,2,\ldots,36\}$. Given a message for encoding, we weight the symbols with weights $1,2,3,\ldots$, beginning with the check digit of the message. Then, the weighted digits are summed together and reduced to the remainder after dividing by 37. Finally, a check symbol is selected such that the sum of the check symbol and the remainder obtained above is congruent to 0 modulo 37.

To calculate this sum of weighted digits easily, a technique called *progressive digiting*, illustrated in Table 2.6, has been developed. In Table 2.6, it is supposed that we want to compute the weighted sum for a message $wxyz$, i.e. $4w+3x+2y+1z$. For each symbol in the message, we first compute the running sum from w to the symbol in question, thereby obtaining the second column in Table 2.6. We can sum these sums again in the same way to obtain the desired weighted sum.

Example 2.6 We assign a distinct number from $\{0,1,2,\ldots,36\}$ to each symbol in the combined alphabet/number set in the following way: "0" $= 0$, "1" $= 1$, "2" $= 2$, ..., "9" $= 9$, "A" $= 10$, "B" $= 11$, "C" $= 12$, ..., "Z" $= 35$, and "space" $= 36$. Then we encode

3B 8.

We proceed with the progressive digiting as shown in Table 2.7 and obtain a weighted sum of 183. Since 183 mod $37 = 35$ and $35+2$ is divisible by 37, it follows that the appended check digit should be

"2" $= 2$.

Table 2.7 *Progressive digiting for the example of "3B 8": we need to add "2" = 2 as a check-digit to make sure that the weighted sum divides 37*

			Sum	Sum of sum
"3"	=	3	3	3
"B"	=	11	14	17
"space"	=	36	50	67
"8"	=	8	58	125
Check-digit	=	??	58	183

$$37 \,\overline{)\,183}\quad \text{quotient } 4;\quad 183 - 148 = 35$$

Table 2.8 *Checking the encoded message "3B 82"*

3	3×5	=	15		
B	11×4	=	44		
"space"	36×3	=	108		
8	8×2	=	16		
2	2×1	=	2		
	Sum	=	185	$= 37 \times 5 \equiv 0$	mod 37

The encoded message is therefore given by

3B 82.

To check whether this is a legitimate message at the receiver, we proceed as shown in Table 2.8.

Now suppose "space" is lost during the transmission such that only "3B82" is received. Such an error can be detected since the weighted sum is now not congruent to 0 mod 37; see Table 2.9. Similarly, the interchange from "82" to

Table 2.9 *Checking the corrupted message "3B82"*

3	3×4	=	12		
B	11×3	=	33		
8	8×2	=	16		
2	2×1	=	2		
	Sum	=	63	$\not\equiv 0$	mod 37

Table 2.10 *Checking the corrupted message "3B 28"*

3	3×5	=	15		
B	11×4	=	44		
"space"	36×3	=	108		
2	2×2	=	4		
8	8×1	=	8		
	Sum	=	179	$\not\equiv 0$	mod 37

"28" can also be detected; see Table 2.10. ◊

In the following we give another two examples of error-detecting codes that are based on modular arithmetic and are widely used in daily commerce.

Example 2.7 The *International Standard Book Number (ISBN)* is usually a 10-digit code used to identify a book uniquely. A typical example of the ISBN is as follows:

$$\underbrace{0}_{\substack{\text{country}\\\text{ID}}} - \underbrace{52}_{\substack{\text{publisher}\\\text{ID}}}\ \underbrace{18-4868}_{\substack{\text{book}\\\text{number}}} - \underbrace{7}_{\substack{\text{check}\\\text{digit}}}$$

where the hyphens do not matter and may appear in different positions. The first digit stands for the country, with 0 meaning the United States and some other English-speaking countries. The next two digits are the publisher ID; here 52 means Cambridge University Press. The next six digits, 18 – 4868, are the publisher-assigned book number. The last digit is the weighted check sum modulo 11 and is represented by "X" if the required check digit is 10.

To confirm that this number is a legitimate ISBN number we proceed as shown in Table 2.11. It checks! ◊

Exercise 2.8 *Check whether 0 – 8044 – 2957 – X is a valid ISBN number.* ◊

Example 2.9 The *Universal Product Code (UPC)* is a 12-digit single parity-check code employed on the bar codes of most merchandise to ensure reliability in scanning. A typical example of UPC takes the form

$$\underbrace{0\ 36000}_{\substack{\text{manufacturer}\\\text{ID}}}\ \underbrace{29145}_{\substack{\text{item}\\\text{number}}}\ \underbrace{2}_{\substack{\text{parity}\\\text{check}}}$$

Table 2.11 *Checking the ISBN number 0 – 5218 – 4868 – 7*

	Sum	Sum of sum
0	0	0
5	5	5
2	7	12
1	8	20
8	16	36
4	20	56
8	28	84
6	34	118
8	42	160
7	49	$209 = 11 \times 19 \equiv 0 \bmod 11$

where the last digit is the parity-check digit. Denote the digits as $x_1, x_2, \ldots, x_{12}$. The parity digit x_{12} is determined such that

$$3(x_1 + x_3 + x_5 + x_7 + x_9 + x_{11}) + (x_2 + x_4 + x_6 + x_8 + x_{10} + x_{12}) \tag{2.46}$$

is a multiple[5] of 10. In this case,

$$3(0+6+0+2+1+5) + (3+0+0+9+4+2) = 60. \tag{2.47}$$

◊

2.7 Trade-off between redundancy and error-detecting capability

As discussed in the previous sections, a single parity check to make the whole message even-parity can help the detection of any single error (or even any odd number of errors). However, if we want to detect the occurrence of more errors in a noisy channel, what can we do for the design of error-detecting codes? Can such a goal be achieved by increasing the number of parity checks, i.e. at the cost of extra redundancy? Fortunately, the answer is positive. Let us consider the following illustrative example.

[5] Note that in this example the modulus 10 is used although this is not a prime. The slightly unusual summation (2.46), however, makes sure that every single error can still be detected. The reason why UPC chooses 10 as the modulus is that the check digit should also range from 0 to 9 so that it can easily be encoded by the bar code.

Example 2.10 For an information source of eight possible outputs, obviously each output can be represented by a binary 3-tuple, say (x_1,x_2,x_3). Suppose three parity checks x_4, x_5, x_6 are now appended to the original message by the following equations:

$$\begin{cases} x_4 = x_1 + x_2, \\ x_5 = x_1 + x_3, \\ x_6 = x_2 + x_3, \end{cases} \tag{2.48}$$

to form a legitimate codeword $(x_1,x_2,x_3,x_4,x_5,x_6)$. Compared with the single parity-check code, this code increases the excess redundancy from $1/3$ to $3/3$. Let $(y_1,y_2,y_3,y_4,y_5,y_6)$ be the received vector as $(x_1,x_2,x_3,x_4,x_5,x_6)$ is transmitted. If at least one of the following parity-check equations is violated:

$$\begin{cases} y_4 = y_1 + y_2, \\ y_5 = y_1 + y_3, \\ y_6 = y_2 + y_3, \end{cases} \tag{2.49}$$

the occurrence of an error is detected.

For instance, consider the case of a single error in the ith position such that

$$y_i = x_i + 1 \quad \text{and} \quad y_\ell = x_\ell, \quad \forall \ell \in \{1,2,\ldots,6\} \setminus \{i\}. \tag{2.50}$$

It follows that

$$\begin{cases} y_4 \neq y_1 + y_2,\ y_5 \neq y_1 + y_3 & \text{if } i = 1, \\ y_4 \neq y_1 + y_2,\ y_6 \neq y_2 + y_3 & \text{if } i = 2, \\ y_5 \neq y_1 + y_3,\ y_6 \neq y_2 + y_3 & \text{if } i = 3, \\ y_4 \neq y_1 + y_2 & \text{if } i = 4, \\ y_5 \neq y_1 + y_3 & \text{if } i = 5, \\ y_6 \neq y_2 + y_3 & \text{if } i = 6. \end{cases} \tag{2.51}$$

Therefore, all single errors can be successfully detected. In addition, consider the case of a double error in the ith and jth positions, respectively, such that

$$y_i = x_i + 1, \quad y_j = x_j + 1, \quad \text{and} \quad y_\ell = x_\ell, \quad \forall \ell \in \{1,2,\ldots,6\} \setminus \{i,j\}. \tag{2.52}$$

We then have

$$
\begin{cases}
y_5 \neq y_1 + y_3,\ y_6 \neq y_2 + y_3 & \text{if } (i,j) = (1,2),\\
y_4 \neq y_1 + y_2,\ y_6 \neq y_2 + y_3 & \text{if } (i,j) = (1,3),\\
y_5 \neq y_1 + y_3 & \text{if } (i,j) = (1,4),\\
y_4 \neq y_1 + y_2 & \text{if } (i,j) = (1,5),\\
y_4 \neq y_1 + y_2,\ y_5 \neq y_1 + y_3,\ y_6 \neq y_2 + y_3 & \text{if } (i,j) = (1,6),\\
y_4 \neq y_1 + y_2,\ y_5 \neq y_1 + y_3 & \text{if } (i,j) = (2,3),\\
y_6 \neq y_2 + y_3 & \text{if } (i,j) = (2,4),\\
y_4 \neq y_1 + y_2,\ y_5 \neq y_1 + y_3,\ y_6 \neq y_2 + y_3 & \text{if } (i,j) = (2,5),\\
y_4 \neq y_1 + y_2 & \text{if } (i,j) = (2,6),\\
y_4 \neq y_1 + y_2,\ y_5 \neq y_1 + y_3,\ y_6 \neq y_2 + y_3 & \text{if } (i,j) = (3,4),\\
y_6 \neq y_2 + y_3 & \text{if } (i,j) = (3,5),\\
y_5 \neq y_1 + y_3 & \text{if } (i,j) = (3,6),\\
y_4 \neq y_1 + y_2,\ y_5 \neq y_1 + y_3 & \text{if } (i,j) = (4,5),\\
y_4 \neq y_1 + y_2,\ y_6 \neq y_2 + y_3 & \text{if } (i,j) = (4,6),\\
y_5 \neq y_1 + y_3,\ y_6 \neq y_2 + y_3 & \text{if } (i,j) = (5,6).
\end{cases}
\tag{2.53}
$$

Hence, this code can detect any pattern of double errors. ◊

Exercise 2.11 *Unfortunately, not all triple errors may be caught by the code of Example 2.10. Can you give an example for verification?* ◊

Without a proper design, however, increasing the number of parity checks may not always improve the error-detecting capability. For example, consider another code which appends the parity checks by

$$
\begin{cases}
x_4 = x_1 + x_2 + x_3,\\
x_5 = x_1 + x_2 + x_3,\\
x_6 = x_1 + x_2 + x_3.
\end{cases}
\tag{2.54}
$$

In this case, x_5 and x_6 are simply repetitions of x_4. Following a similar discussion as in Example 2.10, we can show that all single errors are still detectable. But if the following double error occurs during the transmission:

$$
y_1 = x_1 + 1, \quad y_2 = x_2 + 1, \quad \text{and} \quad y_\ell = x_\ell, \quad \forall 3 \leq \ell \leq 6, \tag{2.55}
$$

none of the three parity-check equations corresponding to (2.54) will be violated. This code thus is not double-error-detecting even though the same amount of redundancy is required as in the code (2.48).

2.8 Further reading

In this chapter simple coding schemes, e.g. single parity-check codes, burst error-detecting codes, and weighted codes, have been introduced to detect the presence of channel errors. However, there exists a class of linear block codes, called *cyclic codes*, which are probably the most widely used form of error-detecting codes. The popularity of cyclic codes arises primarily from the fact that these codes can be implemented with extremely cost-effective electronic circuits. The codes themselves also possess a high degree of structure and regularity (which gives rise to the promising advantage mentioned above), and there is a certain beauty and elegance in the corresponding theory. Interested readers are referred to [MS77], [Wic94], and [LC04] for more details of cyclic codes.

References

[LC04] Shu Lin and Daniel J. Costello, Jr., *Error Control Coding*, 2nd edn. Prentice Hall, Upper Saddle River, NJ, 2004.

[MS77] F. Jessy MacWilliams and Neil J. A. Sloane, *The Theory of Error-Correcting Codes*. North-Holland, Amsterdam, 1977.

[Wic94] Stephen B. Wicker, *Error Control Systems for Digital Communication and Storage*. Prentice Hall, Englewood Cliffs, NJ, 1994.

3
Repetition and Hamming codes

The theory of error-correcting codes comes from the need to protect information from corruption during transmission or storage. Take your CD or DVD as an example. Usually, you might convert your music into MP3 files[1] for storage. The reason for such a conversion is that MP3 files are more compact and take less storage space, i.e. they use fewer binary digits (bits) compared with the original format on CD. Certainly, the price to pay for a smaller file size is that you will suffer some kind of distortion, or, equivalently, losses in audio quality or fidelity. However, such loss is in general indiscernible to human audio perception, and you can hardly notice the subtle differences between the uncompressed and compressed audio signals. The compression of digital data streams such as audio music streams is commonly referred to as *source coding*. We will consider it in more detail in Chapters 4 and 5.

What we are going to discuss in this chapter is the opposite of compression. After converting the music into MP3 files, you might want to store these files on a CD or a DVD for later use. While burning the digital data onto a CD, there is a special mechanism called *error control coding* behind the CD burning process. Why do we need it? Well, the reason is simple. Storing CDs and DVDs inevitably causes small scratches on the disk surface. These scratches impair the disk surface and create some kind of lens effect so that the laser reader might not be able to retrieve the original information correctly. When this happens, the stored files are corrupted and can no longer be used. Since the scratches are inevitable, it makes no sense to ask the users to keep the disks in perfect condition, or discard them once a perfect read-out from the disk becomes impossible. Therefore, it would be better to have some kind of engineering mechanism to protect the data from being compromised by minor scratches.

[1] MP3 stands for *MPEG-2 audio layer 3*, where MPEG is the abbreviation for *moving picture experts group*.

We use error-correcting codes to accomplish this task. Error-correcting codes are also referred to as *channel coding* in general.

First of all, you should note that it is impossible to protect the stored MP3 files from impairment without increasing the file size. To see this, say you have a binary data stream $\mathbf{s}$ of length k bits. If the protection mechanism were not allowed to increase the length, after endowing $\mathbf{s}$ with some protection capability, the resulting stream $\mathbf{x}$ is at best still of length k bits. Then the whole protection process is nothing but a mapping from a k-bit stream to another k-bit stream. Such mapping is, at its best, one-to-one and onto, i.e. a bijection, since if it were not a bijection, it would not be possible to recover the original data. On the other hand, because of the bijection, when the stored data stream $\mathbf{x}$ is corrupted, it is impossible to recover the original $\mathbf{s}$. Therefore, we see that the protection process (henceforth we will refer to it as an *encoding process*) must be an injection, meaning $\mathbf{x}$ must have length larger than k, say n, so that when $\mathbf{x}$ is corrupted, there is a chance that $\mathbf{s}$ may be recovered by using the extra $(n-k)$ bits we have used for storing extra information.

How to encode efficiently a binary stream of length k with minimum $(n-k)$ extra bits added so that the length k stream $\mathbf{s}$ is well protected from corruption is the major concern of error-correcting codes. In this chapter, we will briefly introduce two kinds of error-correcting codes: the *repetition code* and the *Hamming code*. The repetition code, as its name suggests, simply repeats information and is the simplest error-protecting/correcting scheme. The Hamming code, developed by Richard Hamming when he worked at Bell Labs in the late 1940s (we will come back to this story in Section 3.3.1), on the other hand, is a bit more sophisticated than the repetition code. While the original Hamming code is actually not that much more complicated than the repetition code, it turns out to be optimal in terms of sphere packing in some high-dimensional space. Specifically, this means that for certain code length and error-correction capability, the Hamming code actually achieves the maximal possible rate, or, equivalently, it requires the fewest possible extra bits.

Besides error correction and data protection, the Hamming code is also good in many other areas. Readers who wish to know more about these subjects are referred to Chapter 8, where we will briefly discuss two other uses of the Hamming code. We will show in Section 8.1 how the Hamming code relates to a geometric subject called *projective geometry*, and in Section 8.2 how the Hamming code can be used in some mathematical games.

3.1 Arithmetics in the binary field

Prior to introducing the codes, let us first study the arithmetics of binary operations (see also Section 2.1). These are very important because the digital data is binary, i.e. each binary digit is either of value 0 or 1, and the data will be processed in a binary fashion. By binary operations we mean binary addition, subtraction, multiplication, and division. The binary addition is a modulo-2 addition, i.e.

$$\begin{aligned} 0+0&=0,\\ 1+0&=1,\\ 0+1&=1,\\ 1+1&=0. \end{aligned} \tag{3.1}$$

The only difference between binary and usual additions is the case of $1+1$. Usual addition would say $1+1=2$. But since we are working with modulo-2 addition, meaning the sum is taken as the remainder when divided by 2, the remainder of 2 divided by 2 equals 0, hence we have $1+1=0$ in binary arithmetics.

By moving the second operand to the right of these equations, we obtain subtractions:

$$\begin{aligned} 0&=0-0,\\ 1&=1-0,\\ 0&=1-1,\\ 1&=0-1. \end{aligned} \tag{3.2}$$

Further, it is interesting to note that the above equalities also hold if we replace "$-$" by "$+$". Then we realize that, in binary, subtraction is the same as addition. This is because the remainder of -1 divided by 2 equals 1, meaning -1 is considered the same as 1 in binary. In other words,

$$a-b=a+(-1)\times b=a+(1)\times b=a+b. \tag{3.3}$$

Also, it should be noted that the above implies

$$a-b=b-a=a+b \tag{3.4}$$

in binary, while this is certainly false for real numbers.

Multiplication in binary is the same as usual, and we have

$$\begin{aligned} 0 \times 0 &= 0, \\ 1 \times 0 &= 0, \\ 0 \times 1 &= 0, \\ 1 \times 1 &= 1. \end{aligned} \tag{3.5}$$

The same holds also for division.

Exercise 3.1 *Show that the laws of association and distribution hold for binary arithmetics. That is, show that for any $a, b, c \in \{0, 1\}$ we have*

$$\begin{aligned} a+b+c &= (a+b)+c = a+(b+c) && \textit{(additive associative law)}, \\ a\times b\times c &= (a\times b)\times c = a\times(b\times c) && \textit{(multiplicative associative law)}, \\ a\times(b+c) &= (a\times b)+(a\times c) && \textit{(distributive law)}. \end{aligned}$$

◇

Exercise 3.2 *In this chapter, we will use the notation $\stackrel{?}{=}$ to denote a conditional equality, by which we mean that we are unsure whether the equality holds. Show that the condition of $a \stackrel{?}{=} b$ in binary is the same as $a+b \stackrel{?}{=} 0$.* ◇

3.2 Three-times repetition code

A binary digit (or *bit* in short) s is to be stored on CD, but it could be corrupted for some reason during read-out. To recover the corrupted data, a straightforward means of protection is to store as many copies of s as possible. For simplicity, say we store three copies. Such a scheme is called the *three-times repetition code*. Thus, instead of simply storing s, we store (s, s, s). To distinguish them, let us denote the first s as x_1 and the others as x_2 and x_3. In other words, we have

$$\begin{cases} x_2 = x_3 = 0 & \text{if } x_1 = 0, \\ x_2 = x_3 = 1 & \text{if } x_1 = 1, \end{cases} \tag{3.6}$$

and the possible values of (x_1, x_2, x_3) are (000) and (111).

When you read out the stream (x_1, x_2, x_3) from a CD, you must check whether $x_1 = x_2$ and $x_1 = x_3$ in order to detect if there was a data corruption. From Exercise 3.2, this can be achieved by the following computation:

$$\begin{cases} \text{data clean} & \text{if } x_1 + x_2 = 0 \text{ and } x_1 + x_3 = 0, \\ \text{data corrupted} & \text{otherwise.} \end{cases} \tag{3.7}$$

For example, if the read-out is $(x_1, x_2, x_3) = (000)$, then you might say the data is clean. Otherwise, if the read-out shows $(x_1, x_2, x_3) = (001)$ you immediately find $x_1 + x_3 = 1$ and the data is corrupted.

Now say that the probability of writing in 0 and reading out 1 is p, and the same for writing in 1 and reading out 0. You see that a bit is corrupted with probability p and remains clean with probability $(1-p)$. Usually we can assume $p < 1/2$, meaning the data is more likely to be clean than corrupted. In the case of $p > 1/2$, a simple bit-flipping technique of treating the read-out of 1 as 0 and 0 as 1 would do the trick.

Thus, when $p < 1/2$, the only possibilities for data corruption going undetected are the cases when the read-out shows (111) given writing in was (000) and when the read-out shows (000) given writing in was (111). Each occurs with probability[2] $p^3 < 1/8$. Compared with the case when the data is unprotected, the probability of undetectable corruption drops from p to p^3. It means that when the read-out shows either (000) or (111), we are more confident that such a read-out is clean.

The above scheme is commonly referred to as *error detection* (see also Chapter 2), by which we mean we only detect whether the data is corrupted, but we do not attempt to correct the errors. However, our goal was to correct the corrupted data, not just detect it. This can be easily achieved with the repetition code. Consider the case of a read-out (001): you would immediately guess that the original data is more likely to be (000), which corresponds to the binary bit $s = 0$. On the other hand, if the read-out shows (101), you would guess the second bit is corrupted and the data is likely to be (111) and hence determine the original $s = 1$.

There is a good reason for such a guess. Again let us denote by p the probability of a read-out bit being corrupted, and let us assume[3] that the probability of s being 0 is $1/2$ (and of course the same probability for s being 1). Then, given the read-out (001), the probability of the original data being (000) can be computed as follows. Here again we assume that the read-out bits are corrupted independently. Assuming $\Pr[s = 0] = \Pr[s = 1] = 1/2$, it is clear that

$$\Pr(\text{Writing in } (000)) = \Pr(\text{Writing in } (111)) = \frac{1}{2}. \tag{3.8}$$

It is also easy to see that

[2] Here we assume each read-out bit is corrupted independently, meaning whether one bit is corrupted or not has no effect on the other bits being corrupted or not. With this independence assumption, the probability of having three corrupted bits is $p \cdot p \cdot p = p^3$.

[3] Why do we need this assumption? Would the situation be different without this assumption? Take the case of $\Pr[s = 0] = 0$ and $\Pr[s = 1] = 1$ as an example.

$$\begin{aligned}
&\Pr(\text{Writing in } (000) \text{ and reading out } (001)) \\
&\quad = \Pr(\text{Writing in } (000)) \cdot \Pr(\text{Reading out } (001) \mid \text{Writing in } (000)) && (3.9) \\
&\quad = \Pr(\text{Writing in } (000)) \cdot \Pr(0 \to 0) \cdot \Pr(0 \to 0) \cdot \Pr(0 \to 1) && (3.10) \\
&\quad = \frac{1}{2} \cdot (1-p) \cdot (1-p) \cdot p && (3.11) \\
&\quad = \frac{(1-p)^2 p}{2}. && (3.12)
\end{aligned}$$

Similarly, we have

$$\Pr(\text{Writing in } (111) \text{ and reading out } (001)) = \frac{(1-p)p^2}{2}. \tag{3.13}$$

These together show that

$$\begin{aligned}
&\Pr(\text{Reading out } (001)) \\
&\quad = \Pr(\text{Writing in } (000) \text{ and reading out } (001)) \\
&\qquad + \Pr(\text{Writing in } (111) \text{ and reading out } (001)) && (3.14) \\
&\quad = \frac{(1-p)p}{2}. && (3.15)
\end{aligned}$$

Thus

$$\begin{aligned}
&\Pr(\text{Writing in } (000) \mid \text{Reading out } (001)) \\
&\quad = \frac{\Pr(\text{Writing in } (000) \text{ and reading out } (001))}{\Pr(\text{Reading out } (001))} && (3.16) \\
&\quad = 1 - p. && (3.17)
\end{aligned}$$

Similarly, it can be shown that

$$\Pr(\text{Writing in } (111) \mid \text{Reading out } (001)) = p. \tag{3.18}$$

As $p < 1/2$ by assumption, we immediately see that

$$1 - p > p, \tag{3.19}$$

and, given that the read-out is (001), the case of writing in (111) is less likely. Hence we would guess the original data is more likely to be (000) due to its higher probability. Arguing in a similar manner, we can construct a table for decoding, shown in Table 3.1.

From Table 3.1, we see that given the original data being (000), the correctable error events are the ones when the read-outs are (100), (010), and (001), i.e. the ones when only one bit is in error. The same holds for the other write-in of (111). Thus we say that the three-times repetition code is a *single-error-correcting code*, meaning the code is able to correct all possible one-bit errors. If there are at least two out of the three bits in error during read-out,

Table 3.1 *Decoding table for the repetition code based on probability*

Read-outs	Likely original	Decoded output
$(000), (100), (010), (001)$	(000)	$s = 0$
$(111), (011), (101), (110)$	(111)	$s = 1$

then this code is bound to make an erroneous decision as shown in Table 3.1. The probability of having an erroneous decoded output is given by

$$\Pr(\text{Uncorrectable error}) = 3p^2(1-p) + p^3 \tag{3.20}$$

that is smaller than the original p.

Exercise 3.3 *Prove* $3p^2(1-p) + p^3 < p$ *for* $p \in (0, 1/2)$. ◊

Exercise 3.4 *It should be noted that Table 3.1 is obtained under the assumption of* $\Pr[s = 0] = \Pr[s = 1] = 1/2$. *What if* $\Pr[s = 0] = 0$ *and* $\Pr[s = 1] = 1$*? Reconstruct the table for this case and conclude that* $\Pr(\text{Uncorrectable error}) = 0$. *Then rethink whether you need error protection and correction in this case.* ◊

To summarize this section, below we give a formal definition of an error-correcting code.

Definition 3.5 A code $\mathscr{C}$ is said to be an (n, k) *error-correcting code* if it is a scheme of mapping k bits into n bits, and we say $\mathscr{C}$ has code rate $\mathsf{R} = k/n$. We say $\mathscr{C}$ is a *t-error-correcting code* if $\mathscr{C}$ is able to correct any t or fewer errors in the received n-vector. Similarly, we say $\mathscr{C}$ is an *e-error-detecting code* if $\mathscr{C}$ is able to detect any e or fewer errors in the received n-vector.

With the above definition, we see that the three-times repetition code is a $(3, 1)$ error-correcting code with code rate $\mathsf{R} = 1/3$, and it is a 1-error-correcting code. When being used purely for error detection, it is also a 2-error-detecting code. Moreover, in terms of the error correction or error detection capability, we have the following two theorems. The proofs are left as an exercise.

Theorem 3.6 *Let* $\mathscr{C}$ *be an* (n, k) *binary error-correcting code that is t-error-correcting. Then assuming a raw bit error probability of* p, *we have*

$$\begin{aligned}\Pr(\text{Uncorrectable error}) \le \binom{n}{t+1} p^{t+1}(1-p)^{n-t-1} \\ + \binom{n}{t+2} p^{t+2}(1-p)^{n-t-2} + \cdots + \binom{n}{n} p^n,\end{aligned} \tag{3.21}$$

where $\binom{n}{\ell}$ is the binomial coefficient defined as

$$\binom{n}{\ell} \triangleq \frac{n!}{\ell!(n-\ell)!}. \tag{3.22}$$

Theorem 3.7 *Let $\mathscr{C}$ be an (n,k) binary error-correcting code that is e-error-detecting. Then assuming a raw bit error probability of p, we have*

$$\Pr(\text{Undetectable error}) \leq \binom{n}{e+1} p^{e+1}(1-p)^{n-e-1} + \binom{n}{e+2} p^{e+2}(1-p)^{n-e-2} + \cdots + \binom{n}{n} p^{n}. \tag{3.23}$$

Exercise 3.8 *Prove Theorems 3.6 and 3.7.*

Hint: For the situation of Theorem 3.6, if a code is t-error-correcting, we know that it can correctly deal with all error patterns of t or fewer errors. For error patterns with more than t errors, we do not know: some of them might be corrected; some not. Hence, as an upper bound to the error probability, assume that any error pattern with more than t errors cannot be corrected. The same type of thinking also works for Theorem 3.7. ◊

Recall that in (3.6) and (3.7), given the binary bit s, we use $x_1 = x_2 = x_3 = s$ to generate the length-3 binary stream (x_1, x_2, x_3), and use $x_1 + x_2 \stackrel{?}{=} 0$ and $x_1 + x_3 \stackrel{?}{=} 0$ to determine whether the read-out has been corrupted. In general, it is easier to rewrite the two processes using matrices; i.e., we have the following:

$$\begin{pmatrix} x_1 & x_2 & x_3 \end{pmatrix} = s \begin{pmatrix} 1 & 1 & 1 \end{pmatrix} \quad \text{(generating equation)}, \tag{3.24}$$

$$\begin{pmatrix} 1 & 1 & 0 \\ 1 & 0 & 1 \end{pmatrix} \begin{pmatrix} x_1 \\ x_2 \\ x_3 \end{pmatrix} \stackrel{?}{=} \begin{pmatrix} 0 \\ 0 \end{pmatrix} \quad \text{(check equations)}. \tag{3.25}$$

The two matrix equations above mean that we use the matrix

$$\mathsf{G} = \begin{pmatrix} 1 & 1 & 1 \end{pmatrix} \tag{3.26}$$

to generate the length-3 binary stream and use the matrix

$$\mathsf{H} = \begin{pmatrix} 1 & 1 & 0 \\ 1 & 0 & 1 \end{pmatrix} \tag{3.27}$$

to check whether the data is corrupted. Thus, the matrix G is often called the *generator matrix* and H is called the *parity-check matrix*. We have the following definition.

Definition 3.9 Let $\mathscr{C}$ be an (n,k) error-correcting code that maps length-k binary streams $\mathbf{s}$ into length-n binary streams $\mathbf{x}$. We say $\mathscr{C}$ is a *linear code* if there exist a binary matrix G of size $(k \times n)$ and a binary matrix H of size $((n-k) \times n)$ such that the mapping from $\mathbf{s}$ to $\mathbf{x}$ is given by

$$\mathbf{x} = (x_1 \quad \cdots \quad x_n) = \underbrace{(s_1 \quad \cdots \quad s_k)}_{=\mathbf{s}} \mathsf{G} \tag{3.28}$$

and the check equations are formed by

$$\mathsf{H}\mathbf{x}^{\mathsf{T}} \stackrel{?}{=} \begin{pmatrix} 0 \\ \vdots \\ 0 \end{pmatrix}, \tag{3.29}$$

where by $\mathbf{x}^{\mathsf{T}}$ we mean the transpose of vector $\mathbf{x}$ (rewriting horizontal rows as vertical columns and vice versa). The vector $\mathbf{x}$ is called a *codeword* associated with the binary *message* $\mathbf{s}$.

Exercise 3.10 *With linear codes, the detection of corrupted read-outs is extremely easy. Let $\mathscr{C}$ be an (n,k) binary linear code with a parity-check matrix H of size $((n-k) \times n)$. Given any read-out $\mathbf{y} = (y_1, \ldots, y_n)$, show that $\mathbf{y}$ is corrupted if $\mathsf{H}\mathbf{y}^{\mathsf{T}} \neq \mathbf{0}^{\mathsf{T}}$, i.e. if at least one parity-check equation is unsatisfied. It should be noted that the converse is false in general.*[4] ◊

Exercise 3.11 *Let $\mathscr{C}$ be an (n,k) binary linear code with generator matrix G of size $(k \times n)$ and parity-check matrix H of size $((n-k) \times n)$. Show that the product matrix $\mathsf{H}\mathsf{G}^{\mathsf{T}}$ must be a matrix whose entries are either 0 or multiples of 2; hence, after taking modulo reduction by 2, we have $\mathsf{H}\mathsf{G}^{\mathsf{T}} = 0$, an all-zero matrix.* ◊

Exercise 3.12 (Dual code) *In Definition 3.9, we used matrix G to generate the codeword $\mathbf{x}$ given the binary message $\mathbf{s}$ and used the matrix H to check the integrity of read-outs of $\mathbf{x}$ for an (n,k) linear error-correcting code $\mathscr{C}$. On the other hand, it is possible to reverse the roles of G and H. The process is detailed as follows. Given $\mathscr{C}$, G, and H, we define the* dual code *$\mathscr{C}^{\perp}$ of $\mathscr{C}$ by encoding the length-$(n-k)$ binary message $\mathbf{s}'$ as $\mathbf{x}' = \mathbf{s}'\mathsf{H}$ and check the integrity of $\mathbf{x}'$ using $\mathsf{G}\mathbf{x}'^{\mathsf{T}} \stackrel{?}{=} \mathbf{0}^{\mathsf{T}}$. Based on the above, verify the following:*

(a) The dual code $\mathscr{C}^{\perp}$ of the three-times repetition code is a $(3,2)$ linear code with rate $\mathsf{R}' = 2/3$, and

(b) $\mathscr{C}^{\perp}$ is a 1-error-detecting code.

[4] For the *correction* of corrupted read-outs of linear codewords, see the discussion around (3.34).

This code is called the single parity-check code; *see Chapter 2.* ◊

In the above exercise, we have introduced the concept of a dual code. The dual code is useful in the sense that once you have an (n,k) linear code with generator matrix G and parity-check matrix H, you immediately get another $(n,n-k)$ linear code for free, simply by reversing the roles of G and H. However, readers should be warned that this is very often not for the purpose of error correction. Specifically, throughout the studies of various kinds of linear codes, it is often found that if the linear code $\mathscr{C}$ has a very strong error correction capability, then its dual code $\mathscr{C}^{\perp}$ is highly likely to be weak. Conversely, if $\mathscr{C}$ is very weak, then its dual $\mathscr{C}^{\perp}$ might be strong. Nevertheless, the duality between $\mathscr{C}$ and $\mathscr{C}^{\perp}$ can be extremely useful when studying the combinatorial properties of a code, such as packing (see Section 3.3.3), covering (see p. 179), weight enumerations, etc.

3.3 Hamming code

In Section 3.2, we discussed the $(3,1)$ three-times repetition code that is capable of correcting all 1-bit errors or detecting all 2-bit errors. The price for such a capability is that we have to increase the file size by a factor of 3. For example, if you have a file of size 700 MB to be stored on CD and, in order to keep the file from corruption, you use a $(3,1)$ three-times repetition code, a space of 2100 MB, i.e. 2.1 GB, is needed to store the encoded data. This corresponds to almost half of the storage capacity of a DVD!

Therefore, we see that while the three-times repetition code is able to provide some error protection, it is highly inefficient in terms of rate. In general, we would like the rate $\mathsf{R} = k/n$ to be as close to 1 as possible so that wastage of storage space is kept to a minimum.

3.3.1 Some historical background

The problem of finding efficient error-correcting schemes, but of a much smaller scale, bothered Richard Wesley Hamming (1915–1998) while he was employed by Bell Telephone Laboratory (Bell Labs) in the late 1940s. Hamming was a mathematician with a Ph.D. degree from the University of Illinois at Urbana-Champaign in 1942. He was a professor at the University of Louisville during World War II, and left to work on the Manhattan Project in 1945, programming a computer to solve the problem of whether the detonation of an atomic bomb would ignite the atmosphere. In 1946, Hamming went to Bell

Labs and worked on the Bell Model V computer, an electromechanical relay-based machine. At that time, inputs to computers were fed in on punch cards, which would invariably have read errors (the same as CDs or DVDs in computers nowadays). Prior to executing the program on the punch cards, a special device in the Bell Model V computer would check and detect errors. During weekdays, when errors were found, the computer would flash lights so the operators could correct the problem. During after-hours periods and at weekends, when there were no operators, the machine simply terminated the program and moved on to the next job. Thus, during the weekends, Hamming grew increasingly frustrated with having to restart his programs from scratch due to the unreliable card reader.

Over the next few years he worked on the problem of error correction, developing an increasingly powerful array of algorithms. In 1950 he published what is now known as the *Hamming code*, which remains in use in some applications today.

Hamming is best known for the Hamming code he developed in 1950, as well as the Hamming window[5] used in designing digital filters, the Hamming bound related to sphere packing theory (see Section 3.3.3), and the Hamming distance as a measure of distortion in digital signals (see Exercise 3.16). Hamming received the Turing award in 1968 and was elected to the National Academy of Engineering in 1980.

Exercise 3.13 *The error-correcting mechanism used in CDs*[6] *for data protection is another type of error-correcting code called* Reed–Solomon (R-S) code, *developed by Irving S. Reed and Gustave Solomon in 1960. The use of R-S codes as a means of error correction for CDs was suggested by Jack van Lint (1932–2004) while he was employed at Philips Labs in 1979. Two consecutive R-S codes are used in serial in a CD. These two R-S codes operate in bytes (B) instead of bits (1 B $=$ 8 bits). The first R-S code takes in 24 B of raw data and encodes them into a codeword of length 28 B. After this, another mechanism called interleaver would take 28 such encoded codewords, each 28 B long, and then permute the overall $28^2 = 784$ B of data symbols. Finally, the second R-S code will take blocks of 28 B and encode them into blocks of*

[5] The Hamming window was actually not due to Hamming, but to John Tukey (1915–2000), who also rediscovered with James Cooley the famous algorithm of the fast Fourier transform (FFT) that was originally invented by Carl Friedrich Gauss in 1805, but whose importance to modern engineering was not realized by the researchers until 160 years later. The wonderful Cooley–Tukey FFT algorithm is one of the key ingredients of your MP3 players, DVD players, and mobile phones. So, you actually have Gauss to thank for it. Amazing, isn't it?

[6] A similar mechanism is also used in DVDs. We encourage you to visit the webpage of Professor Tom Høholdt at http://www2.mat.dtu.dk/people/T.Hoeholdt/DVD/index.html for an extremely stimulating demonstration.

32 B. Thus, the first R-S code can be regarded as a (28 B, 24 B) linear code and the second as a (32 B, 28 B) linear code. Based on the above, determine the actual size (in megabytes (MB)) of digital information that is stored on a CD if a storage capacity of 720 MB is claimed. Also, what is the overall code rate used on a CD? ◊

3.3.2 Encoding and error correction of the $(7,4)$ Hamming code

The original Hamming code is a $(7,4)$ binary linear code with the following generator and parity-check matrices:

$$\mathsf{G} \triangleq \begin{pmatrix} 1 & 1 & 0 & 1 & 0 & 0 & 0 \\ 0 & 1 & 1 & 0 & 1 & 0 & 0 \\ 1 & 1 & 1 & 0 & 0 & 1 & 0 \\ 1 & 0 & 1 & 0 & 0 & 0 & 1 \end{pmatrix} \quad \text{and} \quad \mathsf{H} = \begin{pmatrix} 1 & 0 & 0 & 1 & 0 & 1 & 1 \\ 0 & 1 & 0 & 1 & 1 & 1 & 0 \\ 0 & 0 & 1 & 0 & 1 & 1 & 1 \end{pmatrix}. \tag{3.30}$$

Specifically, the encoder of the $(7,4)$ Hamming code takes in a message of four bits, say $\mathbf{s} = (s_1, s_2, s_3, s_4)$ and encodes them as a codeword of seven bits, say $\mathbf{x} = (p_1, p_2, p_3, s_1, s_2, s_3, s_4)$, using the following generating equations:

$$\begin{cases} p_1 = s_1 + s_3 + s_4, \\ p_2 = s_1 + s_2 + s_3, \\ p_3 = s_2 + s_3 + s_4. \end{cases} \tag{3.31}$$

Mappings from $\mathbf{s}$ to $\mathbf{x}$ are tabulated in Table 3.2.

Table 3.2 *Codewords of the* $(7,4)$ *Hamming code*

Message	Codeword	Message	Codeword
0000	0000000	1000	1101000
0001	1010001	1001	0111001
0010	1110010	1010	0011010
0011	0100011	1011	1001011
0100	0110100	1100	1011100
0101	1100101	1101	0001101
0110	1000110	1110	0101110
0111	0010111	1111	1111111

There are several ways to memorize the $(7,4)$ Hamming code. The simplest

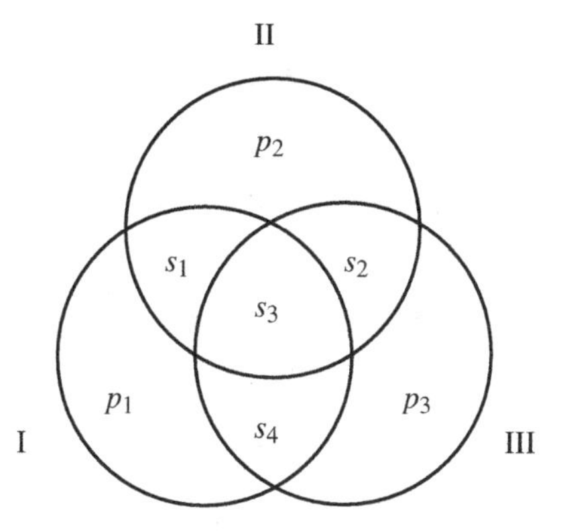

Figure 3.1 Venn diagram of the $(7,4)$ Hamming code.

way is perhaps to use the Venn diagram pointed out by Robert J. McEliece [McE85] and shown in Figure 3.1. There are three overlapping circles: circles I, II, and III. Each circle represents one generating equation as well as one parity-check equation of the $(7,4)$ Hamming code (see (3.31)). For example, circle I corresponds to the first generating equation of $p_1 = s_1 + s_3 + s_4$ and the parity-check equation $p_1 + s_1 + s_3 + s_4 = 0$ since bits s_1, s_3, s_4, and p_1 are included in this circle. Similarly, the second circle, circle II, is for the check equation of $p_2 + s_1 + s_2 + s_3 = 0$, and the third circle III is for the check equation of $p_3 + s_2 + s_3 + s_4 = 0$. Note that each check equation is satisfied if, and only if, there is an even number of 1s in the corresponding circle. Hence, the p_1, p_2, and p_3 are also known as *even parities*.

Using the Venn diagram, error correction of the Hamming code is easy. For example, assume codeword $\mathbf{x} = (0\,1\,1\,0\,1\,0\,0)$ was written onto the CD, but due to an unknown one-bit error the read-out shows $(0\,1\,1\,1\,1\,0\,0)$. We put the read-out into the Venn diagram shown in Figure 3.2. Because of the unknown one-bit error, we see that

- the number of 1s in circle I is 1, an odd number, and a warning (bold circle) is shown;
- the number of 1s in circle II is 3, an odd number, and a warning (bold circle) is shown;
- the number of 1s in circle III is 2, an even number, so no warning (normal circle) is given.

From these three circles we can conclude that the error must not lie in circle III, but lies in both circles I and II. This leaves s_1 as the only possibility, since s_1

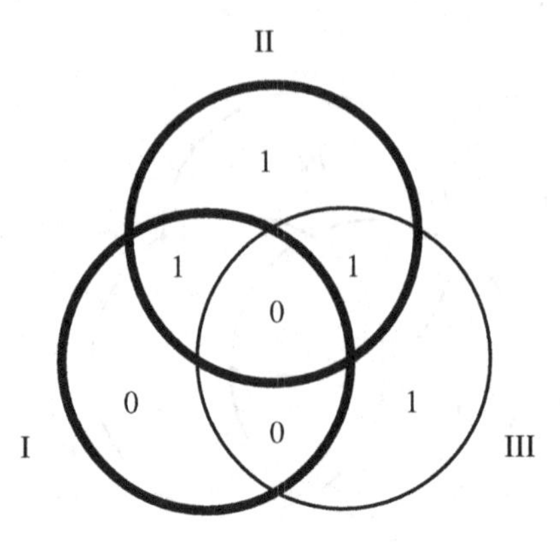

Figure 3.2 Venn diagram used for decoding $(0\,1\,1\,1\,1\,0\,0)$.

is the only point lying in circles I and II but not in III. Hence s_1 must be wrong and should be corrected to a 0 so that both warnings are cleared and all three circles show no warning. This then corrects the erroneous bit as expected.

Let us try another example. What if the read-out is $(1\,1\,1\,0\,1\,0\,0)$? The corresponding Venn diagram is shown in Figure 3.3. Following the same reasoning

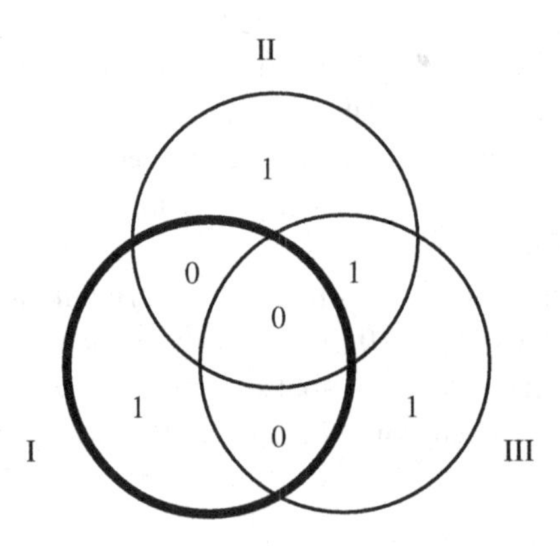

Figure 3.3 Venn diagram used for decoding $(1\,1\,1\,0\,1\,0\,0)$.

as before, we see that the error must lie in circle I, but cannot be in circles II and III. Hence the only possible erroneous bit is p_1. Changing the read-out of p_1 from 1 to 0 will correct the one-bit error.

Now let us revisit the above two error-correcting examples and try to formulate a more systematic approach. In the first example, the read-out was $\mathbf{y} = (0\,1\,1\,1\,1\,0\,0)$. Using the parity-check matrix H defined in (3.30) we see the following connection:

$$\mathsf{H}\mathbf{y}^{\mathsf{T}} = \begin{pmatrix} 1 & 0 & 0 & 1 & 0 & 1 & 1 \\ 0 & 1 & 0 & 1 & 1 & 1 & 0 \\ 0 & 0 & 1 & 0 & 1 & 1 & 1 \end{pmatrix} \begin{pmatrix} 0 \\ 1 \\ 1 \\ 1 \\ 1 \\ 0 \\ 0 \end{pmatrix} = \begin{pmatrix} 1 \\ 1 \\ 0 \end{pmatrix} \iff \begin{pmatrix} \mathbf{I} \\ \mathbf{II} \\ \mathrm{III} \end{pmatrix}. \tag{3.32}$$

This corresponds exactly to the Venn diagram of Figure 3.2. Note that the first entry of $(1\,1\,0)$ corresponds to the first parity-check equation of $p_1 + s_1 + s_3 + s_4 \stackrel{?}{=} 0$ as well as to circle I in the Venn diagram. Since it is unsatisfied, a warning is shown. Similarly for the second and third entries of $(1\,1\,0)$. Now we ask the question: which column of H has the value $(1\,1\,0)^{\mathsf{T}}$? It is the fourth column, which corresponds to the fourth bit of **y**, i.e. s_1 in **x**. Then we conclude that s_1 is in error and should be corrected to a 0. The corrected read-out is therefore $(0\,1\,1\,0\,1\,0\,0)$.

For the second example of a read-out being $\mathbf{y} = (1\,1\,1\,0\,1\,0\,0)$, carrying out the same operations gives

$$\mathsf{H}\mathbf{y}^{\mathsf{T}} = \begin{pmatrix} 1 & 0 & 0 & 1 & 0 & 1 & 1 \\ 0 & 1 & 0 & 1 & 1 & 1 & 0 \\ 0 & 0 & 1 & 0 & 1 & 1 & 1 \end{pmatrix} \begin{pmatrix} 1 \\ 1 \\ 1 \\ 0 \\ 1 \\ 0 \\ 0 \end{pmatrix} = \begin{pmatrix} 1 \\ 0 \\ 0 \end{pmatrix} \iff \begin{pmatrix} \mathbf{I} \\ \mathrm{II} \\ \mathrm{III} \end{pmatrix} \tag{3.33}$$

corresponding to the Venn diagram of Figure 3.3. Since $(1\,0\,0)^{\mathsf{T}}$ is the first column of H, it means the first entry of **y** is in error, and the corrected read-out should be $(0\,1\,1\,0\,1\,0\,0)$.

There is a simple reason why the above error correction technique works. For brevity, let us focus on the first example of $\mathbf{y} = (0\,1\,1\,1\,1\,0\,0)$. This is the case when the fourth bit of **y**, i.e. s_1, is in error. We can write **y** as follows:

$$\mathbf{y} = (0\,1\,1\,1\,1\,0\,0) = \underbrace{(0\,1\,1\,0\,1\,0\,0)}_{=\mathbf{x}} + \underbrace{(0\,0\,0\,1\,0\,0\,0)}_{=\mathbf{e}}, \tag{3.34}$$

where **x** is the original codeword written into a CD and **e** is the *error pattern*.

Recall that $\mathbf{x}$ is a codeword of the $(7,4)$ Hamming code; we must have $\mathsf{H}\mathbf{x}^{\mathsf{T}} = \mathbf{0}^{\mathsf{T}}$ from Definition 3.9. Thus from the distributive law verified in Exercise 3.1 we see that

$$\mathsf{H}\mathbf{y}^{\mathsf{T}} = \mathsf{H}(\mathbf{x}^{\mathsf{T}} + \mathbf{e}^{\mathsf{T}}) = \mathsf{H}\mathbf{x}^{\mathsf{T}} + \mathsf{H}\mathbf{e}^{\mathsf{T}} = \mathbf{0}^{\mathsf{T}} + \mathsf{H}\mathbf{e}^{\mathsf{T}} = \mathsf{H}\mathbf{e}^{\mathsf{T}}. \tag{3.35}$$

Since the only nonzero entry of $\mathbf{e}$ is the fourth entry, left-multiplying $\mathbf{e}^{\mathsf{T}}$ by H gives the fourth column of H. Thus, $\mathsf{H}\mathbf{e}^{\mathsf{T}} = (1\,1\,0)^{\mathsf{T}}$. In other words, we have the following logic deductions: without knowing the error pattern $\mathbf{e}$ in the first place, to correct the one-bit error,

$$\mathsf{H}\mathbf{y}^{\mathsf{T}} = (1\,1\,0)^{\mathsf{T}} \tag{3.36}$$

$$\Longrightarrow \quad \mathsf{H}\mathbf{e}^{\mathsf{T}} = (1\,1\,0)^{\mathsf{T}} \tag{3.37}$$

$$\Longrightarrow \quad \mathbf{e} = (0\,0\,0\,1\,0\,0\,0). \tag{3.38}$$

The last logic deduction relies on the following two facts:

(1) we assume there is only one bit in error, and
(2) all columns of H are distinct.

With the above arguments, we get the following result.

Theorem 3.14 *The $(7,4)$ Hamming code can be classified as one of the following:*

(1) a single-error-correcting code,
(2) a double-error-detecting code.

Proof The proof of the Hamming code being able to correct all one-bit errors follows from the same logic deduction given above. To establish the second claim, simply note that when two errors occur, there are two 1s in the error pattern $\mathbf{e}$, for example $\mathbf{e} = (1\,0\,1\,0\,0\,0\,0)$. Calculating the parity-check equations shows $\mathsf{H}\mathbf{y}^{\mathsf{T}} = \mathsf{H}\mathbf{e}^{\mathsf{T}}$. Note that no two distinct columns of H can be summed to yield $(0\,0\,0)^{\mathsf{T}}$. This means any double-error will give $\mathsf{H}\mathbf{y}^{\mathsf{T}} \neq \mathbf{0}^{\mathsf{T}}$ and hence can be detected. □

From Theorem 3.14 we see that the $(7,4)$ Hamming code is able to correct all one-bit errors. Thus, assuming each bit is in error with probability p, the probability of erroneous correction is given by

$$\Pr(\text{Uncorrectable error}) \leq \binom{7}{2} p^2(1-p)^5 + \binom{7}{3} p^3(1-p)^4 + \cdots + \binom{7}{7} p^7. \tag{3.39}$$

It should be noted that in (3.39) we actually have an equality. This follows from the fact that the Hamming code cannot correct any read-outs having more than one bit in error.

Exercise 3.15 *For error detection of the* $(7,4)$ *Hamming code, recall the check equation* $\mathsf{H}\mathbf{y}^{\mathsf{T}} = \mathsf{H}\mathbf{e}^{\mathsf{T}} \stackrel{?}{=} \mathbf{0}^{\mathsf{T}}$. *Using this relation, first show that an error pattern* $\mathbf{e}$ *is undetectable if, and only if,* $\mathbf{e}$ *is a nonzero codeword. Thus the* $(7,4)$ *Hamming code can detect some error patterns that have more than two errors. Use this fact to show that the probability of a detection error of the* $(7,4)$ *Hamming code is*

$$\Pr(\text{Undetectable error}) = 7p^3(1-p)^4 + 7p^4(1-p)^3 + p^7, \tag{3.40}$$

which is better than what has been claimed by Theorem 3.7.

Hint: Note from Table 3.2 that, apart from the all-zero codeword, there are seven codewords containing three 1s, another seven codewords containing four 1s, and one codeword consisting of seven 1s. ◊

Next we compare the performance of the three-times repetition code with the performance of the $(7,4)$ Hamming code. First, note that both codes are able to correct all one-bit errors or detect all double-bit errors. Yet, the repetition code requires to triple the size of the original file for storage, while the $(7,4)$ Hamming code only needs $7/4 = 1.75$ times the original space. Therefore, the $(7,4)$ Hamming code is more efficient than the three-times repetition code in terms of required storage space.

Before concluding this section, we use the following exercise problem as a quick introduction to another contribution of Hamming. It is related to the topic of *sphere packing*, which will be discussed in Section 3.3.3.

Exercise 3.16 (Hamming distance) *Another contribution of Richard Hamming is the notion of* Hamming distance. *Given any two codewords* $\mathbf{x} = (x_1, x_2, \ldots, x_7)$ *and* $\mathbf{x}' = (x'_1, x'_2, \ldots, x'_7)$, *the Hamming distance between* $\mathbf{x}$ *and* $\mathbf{x}'$ *is the number of places* $\mathbf{x}$ *differs from* $\mathbf{x}'$. *For example, the Hamming distance between* $(1\,0\,1\,1\,1\,0\,0)$ *and* $(0\,1\,1\,1\,0\,0\,1)$ *is 4 since* $\mathbf{x}$ *differs from* $\mathbf{x}'$ *in the first, second, fifth, and seventh positions. Equivalently, you can compute* $(1\,0\,1\,1\,1\,0\,0) + (0\,1\,1\,1\,0\,0\,1) = (1\,1\,0\,0\,1\,0\,1)$. *Then the condition given in Exercise 3.2, namely,* $x_\ell \stackrel{?}{=} x'_\ell$ *is equivalent to* $x_\ell + x'_\ell \stackrel{?}{=} 0$ *for* $\ell = 1, 2, \ldots, 7$, *shows the distance is 4 since there are four 1s appearing in the sum* $(1\,1\,0\,0\,1\,0\,1)$. *Note that the number of ones in a binary vector is called the* Hamming weight *of the vector.*

Now, using Table 3.2, show that every two distinct codewords of a $(7,4)$ *Hamming code are separated by Hamming distance* ≥ 3. ◊

Some of you might wonder why not simply stick to the general definition of Euclidean distance and try to avoid the need for this new definition of Hamming distance. There is a good reason for this. Recall that the Euclidean distance between two distinct points (x,y) and (x',y') is defined as follows:

$$d \triangleq \sqrt{(x-x')^2+(y-y')^2}. \tag{3.41}$$

However, this definition will not work in the binary space. To see this, consider the example of $(x,y) = (0\,0)$ and $(x',y') = (1\,1)$. The Euclidean distance between these two points is given by

$$d = \sqrt{(0-1)^2+(0-1)^2} = \sqrt{1+1} = \sqrt{0} = 0, \tag{3.42}$$

where you should note that $1+1=0$ in binary. Thus, the Euclidean distance fails in the binary space.

3.3.3 Hamming bound: sphere packing

In Exercise 3.16 we have seen that every distinct pair of codewords of the $(7,4)$ Hamming code is separated by Hamming distance at least $d=3$. Thus in terms of geometry we have a picture as shown in Figure 3.4(a). Now if we draw two spheres as shown in Figure 3.4(b) (you might want to think of them as high-dimensional balls), each with radius $\mathrm{R}=1$, centered at $\mathbf{x}$ and $\mathbf{x}'$, respectively, these two spheres would not overlap and must be well-separated. Points within the $\mathbf{x}$-sphere represent the read-outs that are at a distance of at most 1 from $\mathbf{x}$. In other words, the points within the $\mathbf{x}$-sphere are either $\mathbf{x}$ or $\mathbf{x}$ with a one-bit error.

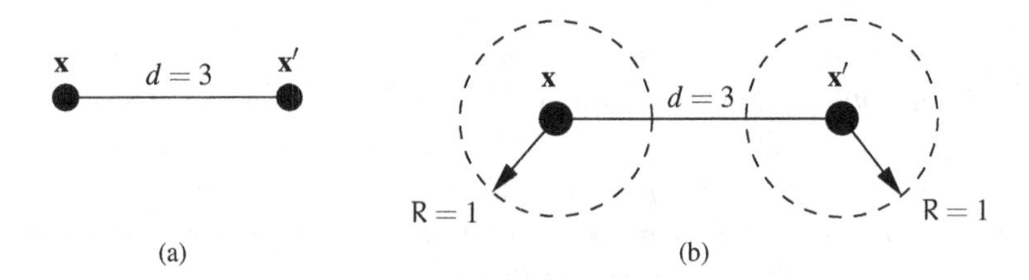

Figure 3.4 Geometry of $\mathbf{x} \neq \mathbf{x}'$ in the $(7,4)$ Hamming code with Hamming distance 3.

If we draw a sphere with radius $\mathrm{R}=1$ centered at each codeword of the $(7,4)$ Hamming code, there will be 16 nonoverlapping spheres since there are 16 codewords and every pair of distinct codewords is separated by a distance of at least 3. Given that a codeword $\mathbf{x}$ was written into the CD, for example, the one-bit error read-out must be at distance 1 from $\mathbf{x}$ and therefore must lie within the radius-1 $\mathbf{x}$-sphere centered at $\mathbf{x}$. It cannot lie in other spheres since the spheres are well-separated. This shows that this one-bit error read-out is closer to $\mathbf{x}$ than to any other codewords of the $(7,4)$ Hamming code.

Thus, correcting a one-bit error is always possible for the $(7,4)$ Hamming code. This can be seen as a geometrical explanation of the single-error-correction capability of the $(7,4)$ Hamming code. We may generalize the above argument slightly and give the following theorem.

Theorem 3.17 *Let $\mathscr{C}$ be an (n,k) error-correcting code (not necessarily linear, i.e. it does not necessarily have the generator and parity-check matrices* G *and* H*). Assume that every distinct pair of codewords in $\mathscr{C}$ is separated by Hamming distance at least d; then $\mathscr{C}$ is a t-error-correcting code with*

$$t = \left\lfloor \frac{d-1}{2} \right\rfloor, \tag{3.43}$$

where by $\lfloor \xi \rfloor$ we mean[7] *the largest integer not larger than ξ. Also, if $\mathscr{C}$ is used only for error detection, then $\mathscr{C}$ is an $(d-1)$-error-detecting code.*

Proof Given d, we can draw spheres with radius t centered at the codewords of $\mathscr{C}$. Since $2t < d$, the spheres must be nonoverlapping. Extending the proof to error detection is obvious. □

The above theorem says that in order to correct more errors, the codewords should be placed as far apart as possible. But this is not what we are interested in here. Instead, we are interested in the reverse direction. We ask the following question.

> *Consider a t-error-correcting code $\mathscr{C}$ that maps input messages to a binary stream of length n. So, we draw spheres of radius t centered at the codewords of $\mathscr{C}$. The spheres do not overlap with each other. What is the maximal number of codewords $\mathscr{C}$ can have? In other words, we are interested in knowing how many nonoverlapping radius-t spheres can be packed into an n-dimensional binary space.*

This is the *sphere packing* problem in discrete mathematics. Let us work through some examples in order to understand the question better.

Example 3.18 The $(7,4)$ Hamming code has 16 codewords, hence 16 spheres with radius 1, since the code is 1-error-correcting.

- The codewords have length 7, with a binary value in each coordinate. So,

[7] For example, $\lfloor 1.1 \rfloor = \lfloor 1.9 \rfloor = 1$.

the number of possible length-7 binary tuples is $2^7 = 128$, meaning there are 128 points in this 7-dimensional binary space.

- Each codeword of the Hamming code is surrounded by a sphere with radius 1. There are $1 + \binom{7}{1} = 8$ points in each sphere. This first "1" corresponds to the center, i.e. distance 0. The remaining $\binom{7}{1}$ points are the ones at distance 1 from the center, i.e. one-bit errors from the codeword.

Thus, the 16 nonoverlapping spheres actually cover $16 \times 8 = 128$ points, which are all the points in the 7-dimensional binary space. We see that the $(7,4)$ Hamming code has the tightest possible packing of radius-1 spheres in the 7-dimensional binary space. ◊

Example 3.19 Let us consider the dual of the $(7,4)$ Hamming code $\mathscr{C}$ in this example. Recall from Exercise 3.12 that the dual code $\mathscr{C}^\perp$ is obtained by reversing the roles of the generating matrix G and the parity-check matrix H of $\mathscr{C}$. That is, we use the parity-check matrix H for encoding and the generator matrix G for checking. Thus the generator matrix $\mathsf{G}^\perp$ of $\mathscr{C}^\perp$ is given by

$$\mathsf{G}^\perp = \mathsf{H} = \begin{pmatrix} 1 & 0 & 0 & 1 & 0 & 1 & 1 \\ 0 & 1 & 0 & 1 & 1 & 1 & 0 \\ 0 & 0 & 1 & 0 & 1 & 1 & 1 \end{pmatrix} \tag{3.44}$$

and it maps binary messages of length 3 to codewords of length 7. All the eight possible codewords are tabulated in Table 3.3.

Table 3.3 *Codewords of the dual code of the* $(7,4)$ *Hamming code*

Message	Codeword	Message	Codeword
0 0 0	0 0 0 0 0 0 0	1 0 0	1 0 0 1 0 1 1
0 0 1	0 0 1 0 1 1 1	1 0 1	1 0 1 1 1 0 0
0 1 0	0 1 0 1 1 1 0	1 1 0	1 1 0 0 1 0 1
0 1 1	0 1 1 1 0 0 1	1 1 1	1 1 1 0 0 1 0

You can check that the codewords are separated by Hamming distance 4 exactly. Hence $\mathscr{C}^\perp$ is able to correct errors up to $t = \left\lfloor \frac{4-1}{2} \right\rfloor = 1$, which is the same as $\mathscr{C}$ does. Thus, it is clear that $\mathscr{C}^\perp$ is not a good packing of radius-1 spheres in the 7-dimensional binary space since it packs only eight spheres, while $\mathscr{C}$ can pack 16 spheres into the same space. ◊

Why are we interested in the packing of radius-t spheres in an n-dimensional space? The reason is simple. Without knowing the parameter k in the first place, i.e. without knowing how many distinct 2^k binary messages can be encoded, by fixing t we make sure that the codewords are immune to errors with at most t error bits. Choosing n means the codewords will be stored in n bits. Being able to pack more radius-t spheres into the n-dimensional spaces means we can have more codewords, hence larger k. This gives a general bound on k, known as the *sphere bound*, and stated in the following theorem.

Theorem 3.20 (Sphere bound) *Let n, k, and t be defined as above. Then we have*

$$2^k \le \frac{2^n}{\binom{n}{0} + \binom{n}{1} + \cdots + \binom{n}{t}}. \tag{3.45}$$

Codes with parameter n, k, and t that achieve equality in (3.45) are called perfect, *meaning a perfect packing.*

Proof Note that 2^k is the number of codewords, while 2^n is the number of points in an n-dimensional binary space, i.e. the number of distinct binary n-tuples. The denominator shows the number of points within a radius-t sphere. The inequality follows from the fact that for t-error-correcting codes the spheres must be nonoverlapping. □

Finally we conclude this section with the following very deep result.

Theorem 3.21 *The only parameters satisfying the bound (3.45) with equality are*

$$\begin{cases} n = 2^u - 1, & k = 2^u - u - 1, \quad t = 1, \quad \text{for any positive integer } u; \\ n = 23, & k = 12, \quad t = 3; \\ n = 2u + 1, & k = 1, \quad t = u, \quad \text{for any positive integer } u. \end{cases} \tag{3.46}$$

This theorem was proven by Aimo Tietäväinen [Tie73] in 1973 after much work by Jack van Lint. One code satisfying the second case of $n = 23$, $k = 12$, and $t = 3$ is the *Golay code*, hand-constructed by Marcel J. E. Golay in 1949 [Gol49].[8] Vera Pless [Ple68] later proved that the Golay code is the only code with these parameters that satisfies (3.45) with equality. The first case is a general Hamming code of order u (see Exercise 3.22 below), and the last case is the $(2u + 1)$-times repetition code, i.e. repeating the message $(2u + 1)$ times.

[8] This paper is only half a page long, but belongs to the most important paper in information theory ever written! Not only did it present the perfect Golay code, but it also gave the generalization of the Hamming code and the first publication of a parity-check matrix. And even though it took over 20 years to prove it, Golay already claimed in that paper that there were no other perfect codes. For more details on the life of Marcel Golay, see http://www.isiweb.ee.ethz.ch/archive/massey_pub/pdf/BI953.pdf.

Exercise 3.22 (Hamming code of order u) *Recall that the* $(7,4)$ *Hamming code is defined by its parity-check matrix*

$$\mathsf{H} = \begin{pmatrix} 1 & 0 & 0 & 1 & 0 & 1 & 1 \\ 0 & 1 & 0 & 1 & 1 & 1 & 0 \\ 0 & 0 & 1 & 0 & 1 & 1 & 1 \end{pmatrix}. \tag{3.47}$$

Note that the columns of the above (3×7) *matrix consist of all possible nonzero length-3 binary vectors. From this, we can easily define a general Hamming code* $\mathscr{C}_u$ *of order u. Let* H_u *be the matrix whose columns consist of all possible nonzero length-u binary vectors;* H_u *is of size* $(u \times (2^u - 1))$*. Then a general Hamming code* $\mathscr{C}_u$ *is the code defined by the parity-check matrix* H_u *with* $n = 2^u - 1$ *and* $k = 2^u - u - 1$*. Show that*

(a) $\mathscr{C}_u$ *is 2-error-detecting (Hint: Show that* $\mathsf{H}_u\mathbf{e}^\mathsf{T} \neq \mathbf{0}^\mathsf{T}$ *for all nonzero vectors* $\mathbf{e}$ *that have at most two nonzero entries);*
(b) $\mathscr{C}_u$ *is 1-error-correcting (Hint: Show that* $\mathsf{H}_u\mathbf{y}^\mathsf{T} = \mathbf{0}^\mathsf{T}$ *for some nonzero vector* $\mathbf{y}$ *if, and only if,* $\mathbf{y}$ *has at least three nonzero entries. Then use this to conclude that every pair of distinct codewords is separated by Hamming distance at least 3).* ◊

3.4 Further reading

In this chapter we have briefly introduced the theory of error-correcting codes and have carefully studied two example codes: the three-times repetition code and the $(7,4)$ Hamming code. Besides their error correction capabilities, we have also briefly studied the connections of these codes to sphere packing in high-dimensional spaces.

For readers who are interested in learning more about other kinds of error-correcting codes and their practical uses, [Wic94] is an easy place to start, where you can learn about a more general treatment of the Hamming codes. Another book, by Shu Lin and Daniel Costello [LC04], is a comprehensive collection of all modern coding schemes. An old book by Jessy MacWilliams and Neil Sloane [MS77] is the most authentic source for learning the theory of error-correcting codes, but it requires a solid background in mathematics at graduate level.

The Hamming codes are closely related to combinatorial designs, difference sets, and Steiner systems. All are extremely fascinating objects in combinatorics. The interested readers are referred to [vLW01] by Jack van Lint and Richard Wilson for further reading on these subjects. These combinatorial objects are also used in the designs of radar systems, spread spectrum-based cellular communications, and optical fiber communication systems.

The topic of sphere packing is always hard, yet fascinating. Problems therein have been investigated for more than 2000 years, and many remain open. A general discussion of this topic can be found in [CS99]. As already seen in Theorem 3.21, the sphere packing bound is not achievable in almost all cases. Some bounds that are tighter than the sphere packing bound, such as the Gilbert–Varshamov bound, the Plotkin bound, etc., can be found in [MS77] and [Wic94]. In [MS77] a table is provided that lists all known best packings in various dimensions. An updated version can be found in [HP98]. So far, the tightest lower bound on the existence of the densest possible packings is the Tsfasman–Vlăduţ–Zink (TVZ) bound, and there are algebraic geometry codes constructed from function fields defined by the Garcia–Stichtentoth curve that perform better than the TVZ bound, i.e. much denser sphere packings. A good overview of this subject can be found in [HP98].

References

[CS99] John Conway and Neil J. A. Sloane, *Sphere Packings, Lattices and Groups*, 3rd edn. Springer Verlag, New York, 1999.

[Gol49] Marcel J. E. Golay, "Notes on digital coding," *Proceedings of the IRE*, vol. 37, p. 657, June 1949.

[HP98] W. Cary Huffman and Vera Pless, eds., *Handbook of Coding Theory*. North-Holland, Amsterdam, 1998.

[LC04] Shu Lin and Daniel J. Costello, Jr., *Error Control Coding*, 2nd edn. Prentice Hall, Upper Saddle River, NJ, 2004.

[McE85] Robert J. McEliece, "The reliability of computer memories," *Scientific American*, vol. 252, no. 1, pp. 68–73, 1985.

[MS77] F. Jessy MacWilliams and Neil J. A. Sloane, *The Theory of Error-Correcting Codes*. North-Holland, Amsterdam, 1977.

[Ple68] Vera Pless, "On the uniqueness of the Golay codes," *Journal on Combination Theory*, vol. 5, pp. 215–228, 1968.

[Tie73] Aimo Tietäväinen, "On the nonexistence of perfect codes over finite fields," *SIAM Journal on Applied Mathematics*, vol. 24, no. 1, pp. 88–96, January 1973.

[vLW01] Jacobus H. van Lint and Richard M. Wilson, *A Course in Combinatorics*, 2nd edn. Cambridge University Press, Cambridge, 2001.

[Wic94] Stephen B. Wicker, *Error Control Systems for Digital Communication and Storage*. Prentice Hall, Englewood Cliffs, NJ, 1994.

4

Data compression: efficient coding of a random message

In this chapter we will consider a new type of coding. So far we have concentrated on codes that can help detect or even correct errors; we now would like to use codes to represent some information more *efficiently*, i.e. we try to represent the same information using fewer digits on average. Hence, instead of protecting data from errors, we try to *compress* it such as to use less storage space.

To achieve such a compression, we will assume that we know the probability distribution of the messages being sent. If some symbols are more probable than others, we can then take advantage of this by assigning shorter codewords to the more frequent symbols and longer codewords to the rare symbols. Hence, we see that such a code has codewords that are not of fixed length.

Unfortunately, variable-length codes bring with them a fundamental problem: at the receiving end, how do you recognize the end of one codeword and the beginning of the next? To attain a better understanding of this question and to learn more about how to design a good code with a short average codeword length, we start with a motivating example.

4.1 A motivating example

You would like to set up your own telephone system that connects you to your three best friends. The question is how to design efficient binary phone numbers. In Table 4.1 you find six different ways of how you could choose them.

Note that in this example the phone number is a *codeword* for the person we want to talk to. The set of all phone numbers is called *code*. We also assume that you have different probabilities when calling your friends: Bob is your best friend whom you will call in 50% of the times. Alice and Carol are contacted with a frequency of 25% each.

Table 4.1 *Binary phone numbers for a telephone system with three friends*

Friend	Probability	Phone number					
Alice	1/4	0011	001101	0	00	0	10
Bob	1/2	0011	001110	1	11	11	0
Carol	1/4	1100	110000	10	10	10	11
		(i)	(ii)	(iii)	(iv)	(v)	(vi)

Let us discuss the different designs in Table 4.1.

(i) In this design, Alice and Bob have the same phone number. The system obviously will not be able to connect properly.

(ii) This is much better, i.e. the code will actually work. However, the phone numbers are quite long and therefore the design is rather inefficient.

(iii) Now we have a code that is much shorter and, at the same time, we have made sure that we do not use the same codeword twice. However, a closer look reveals that the system will not work. The problem here is that this code is not *uniquely decodable*: if you dial 10 this could mean "Carol" or also "Bob, Alice." Or, in other words, the telephone system will never connect you to Carol, because once you dial 1, it will immediately connect you to Bob.

(iv) This is the first quite efficient code that is functional. But we note something: when calling Alice, why do we have to dial two zeros? After the first zero it is already clear to whom we would like to be connected! Let us fix that in design (v).

(v) This is still uniquely decodable and obviously more efficient than (iv). Is it the most efficient code? No! Since Bob is called most often, he should be assigned the shortest codeword!

(vi) This is the optimal code. Note one interesting property: even though the numbers do not all have the same length, once you finish dialing any of the three numbers, the system immediately knows that you have finished dialing. This is because no codeword is the *prefix*[1] of any other codeword, i.e. it never happens that the first few digits of one codeword are identical to another codeword. Such a code is called *prefix-free* (see Section 4.2). Note that (iii) was not prefix-free: 1 is a prefix of 10.

[1] According to the *Oxford English Dictionary*, a *prefix* is a word, letter, or number placed before another.

From this example we learn the following requirements that we impose on our code design.

- A code needs to be *uniquely decodable.*
- A code should be short; i.e., we want to *minimize the average codeword length* L_{av}, which is defined as follows:

$$\mathsf{L}_{\text{av}} \triangleq \sum_{i=1}^{r} p_i l_i. \tag{4.1}$$

  Here p_i denotes the probability that the source emits the ith symbol, i.e. the probability that the ith codeword $\mathbf{c}_i$ is selected; l_i is the length of the ith codeword $\mathbf{c}_i$; and r is the number of codewords.
- We additionally require the code to be *prefix-free*. Note that this requirement is not necessary, but only convenient. However, we will later see that we lose nothing by asking for it.

Note that any prefix-free code is implicitly uniquely decodable, but not vice versa. We will discuss this issue in more detail in Section 4.7.

4.2 Prefix-free or instantaneous codes

Consider the following code with four codewords:

$$\begin{aligned} \mathbf{c}_1 &= 0 \\ \mathbf{c}_2 &= 10 \\ \mathbf{c}_3 &= 110 \\ \mathbf{c}_4 &= 111 \end{aligned} \tag{4.2}$$

Note that the zero serves as a kind of "comma": whenever we receive a zero (or the code has reached length 3), we know that the codeword has finished. However, this comma still contains useful information about the message as there is still one codeword without it! This is another example of a prefix-free code. We recall the following definition.

Definition 4.1 A code is called *prefix-free* (sometimes also called *instantaneous*) if no codeword is the prefix of another codeword.

The name *instantaneous* is motivated by the fact that for a prefix-free code we can decode instantaneously once we have received a codeword and do not need to wait for later until the decoding becomes unique. Unfortunately, in the literature one also finds that people call a prefix-free code a *prefix code*. This name is confusing because rather than having prefixes it is the point of the code to have *no* prefix! We will stick to the name of *prefix-free codes*.

Consider next the following example:

$$\begin{aligned} \mathbf{c}_1 &= 0 \\ \mathbf{c}_2 &= 01 \\ \mathbf{c}_3 &= 011 \\ \mathbf{c}_4 &= 111 \end{aligned} \tag{4.3}$$

This code is not prefix-free (0 is a prefix of 01 and 011; 01 is a prefix of 011), but it is still uniquely decodable.

Exercise 4.2 *Given the code in (4.3), split the sequence* 0011011110 *into codewords.* ◊

Note the drawback of the code design in (4.3): the receiver needs to wait and see how the sequence continues before it can make a unique decision about the decoding. The code is *not instantaneously* decodable.

Apart from the fact that they can be decoded instantaneously, another nice property of prefix-free codes is that they can very easily be represented by leaves of *decision trees*. To understand this we will next make a small detour and talk about trees and their relation to codes.

4.3 Trees and codes

The following definition is quite straightforward.

Definition 4.3 (Trees) A *rooted tree* consists of a root with some branches, nodes, and leaves, as shown in Figure 4.1. A *binary tree* is a rooted tree in which each node (hence also the root) has exactly two children,[2] i.e. two branches stemming forward.

The clue to this section is to note that *any* binary code can be represented as a binary tree. Simply take any codeword and regard each bit as a decision

[2] The alert reader might wonder why we put so much emphasis on having *exactly* two children. It is quite obvious that if a parent node only had one child, then this node would be useless and the child could be moved back and replace its parent. The reason for our definition, however, has nothing to do with efficiency, but is related to the generalization to D-ary trees where every node has exactly D children. We do not cover such trees in this book; the interested reader is referred to, e.g., [Mas96].

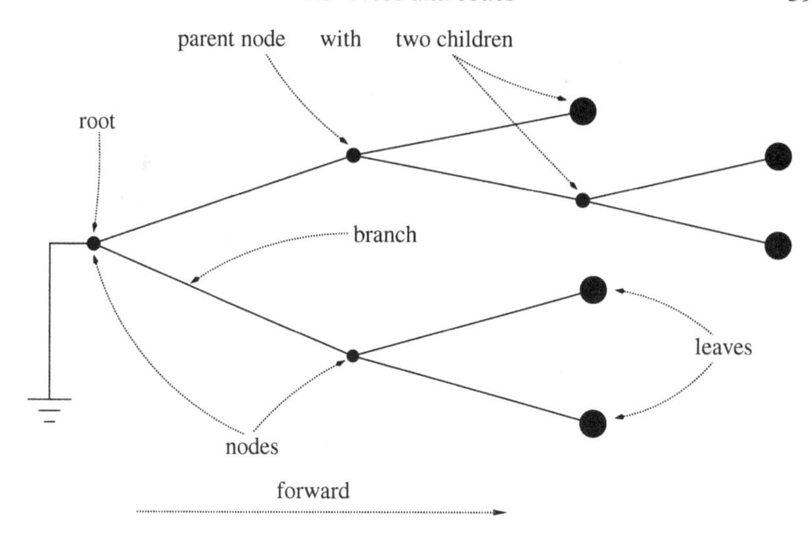

Figure 4.1 A rooted tree (in this case a binary tree) with a *root* (the node that is grounded), four *nodes* (including the root), and five *leaves*. Note that in this book we will always clearly distinguish between *nodes* and *leaves*: a node always has children, while a leaf always is an "end-point" in the tree.

whether to go up ("0") or down[3] ("1"). Hence, every codeword can be represented by a particular path traversing through the tree. As an example, Figure 4.2 shows the binary tree of a binary code with five codewords. Note that on purpose we also keep branches that are not used in order to make sure that the tree is binary.

In Figure 4.3, we show the tree describing the prefix-free code given in (4.2). Note that here every codeword is a leaf. This is no accident.

Lemma 4.4 *A binary code $\{\mathbf{c}_1, \ldots, \mathbf{c}_r\}$ is prefix-free if, and only if, in its binary tree every codeword is a leaf. (But not every leaf necessarily is a codeword; see, e.g., code (iv) in Figure 4.4.)*

Exercise 4.5 *Prove Lemma 4.4.*

Hint: Think carefully about the definition of prefix-free codes (see Definition 4.1). ◊

As mentioned, the binary tree of a prefix-free code might contain leaves that are not codewords. Such leaves are called *unused leaves*.

Some more examples of trees of prefix-free and non-prefix-free codes are shown in Figure 4.4.

[3] It actually does not matter whether 1 means up and 0 down, or vice versa.

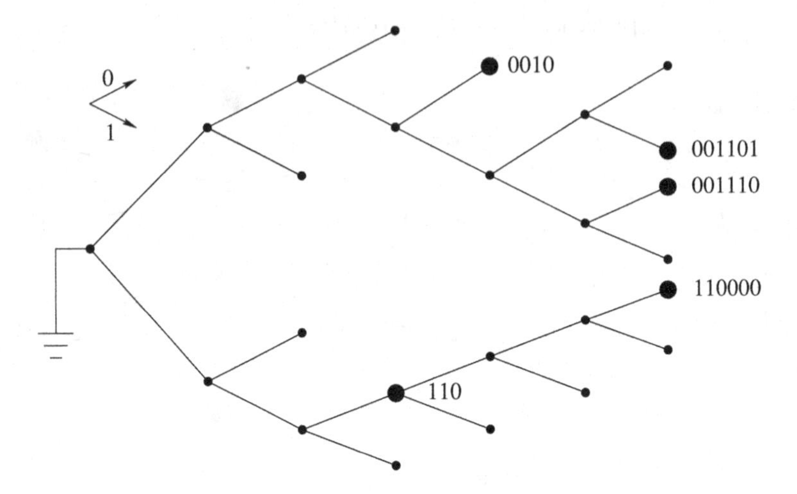

Figure 4.2 An example of a binary tree with five codewords: 110, 0010, 001101, 001110, and 110000. At every node, going upwards corresponds to a 0, and going downwards corresponds to a 1. The node with the ground symbol is the *root* of the tree indicating the starting point.

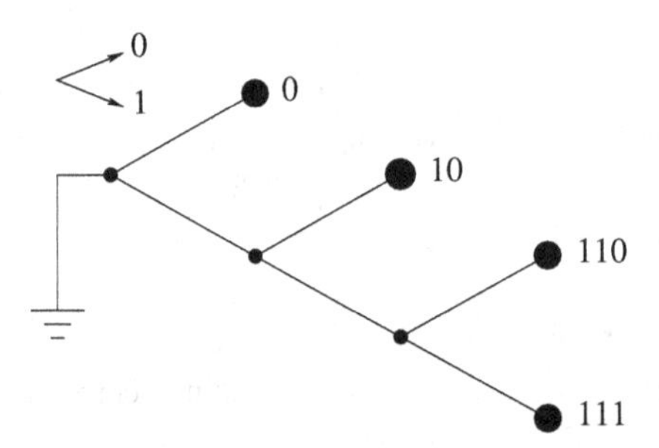

Figure 4.3 Decision tree corresponding to the prefix-free code given in (4.2).

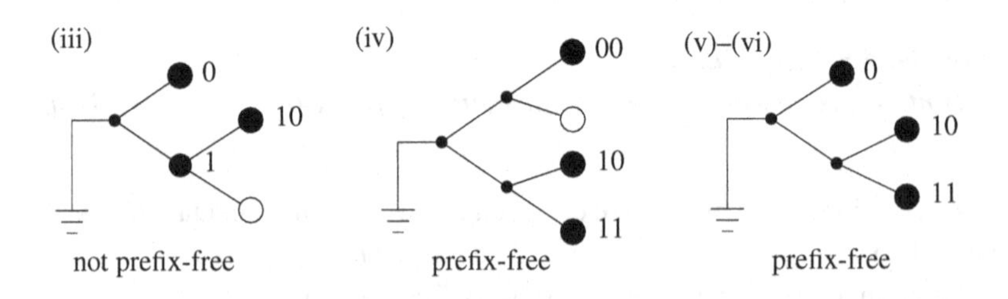

Figure 4.4 Examples of codes and their corresponding trees. The examples are taken from Table 4.1. The prefix-free code (iv) has one unused leaf.

An important concept of trees is the *depths* of their leaves.

Definition 4.6 The *depth of a leaf* in a binary tree is the number of steps it takes when walking from the root forward to the leaf.

As an example, consider again Figure 4.4. Tree (iv) has four leaves, all of them at depth 2. Both tree (iii) and tree (v)–(vi) have three leaves, one at depth 1 and two at depth 2.

We now will derive some interesting properties of trees. Since codes can be represented as trees, we will then be able to apply these properties directly to codes.

Lemma 4.7 (Leaf-Counting and Leaf-Depth Lemma) *The number of leaves n and their depths $l_1, l_2, \ldots, l_n$ in a binary tree satisfy:*

$$n = 1 + \mathrm{N}, \tag{4.4}$$

$$\sum_{i=1}^{n} 2^{-l_i} = 1, \tag{4.5}$$

where N *is the number of nodes (including the root).*

Proof By *extending a leaf* we mean changing a leaf into a node by adding two branches that stem forward. In that process

- we reduce the number of leaves by 1,
- we increase the number of nodes by 1, and
- we increase the number of leaves by 2,

i.e. in total we gain one node and one leaf. This process is depicted graphically in Figure 4.5.

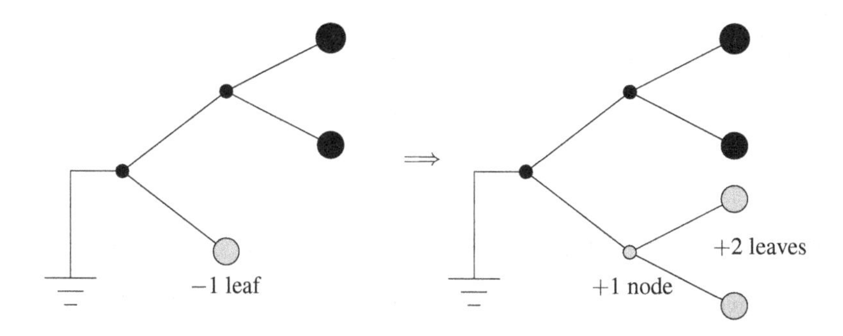

Figure 4.5 Extending a leaf: both the total number of nodes and the total number of leaves is increased by 1.

To prove the first statement (4.4), we start with the *extended root*; i.e., at the beginning we have the root and $n = 2$ leaves. In this case we have $\mathrm{N} = 1$ and (4.4) is satisfied. Now we can grow any tree by continuously extending some leaf, every time increasing the number of leaves and nodes by one each. We see that (4.4) remains valid. By induction this proves the first statement.

We will prove the second statement (4.5) also by induction. We again start with the extended root.

(1) An extended root has two leaves, all at depth 1: $l_i = 1$. Hence,

$$\sum_{i=1}^{n} 2^{-l_i} = \sum_{i=1}^{2} 2^{-1} = 2 \cdot 2^{-1} = 1; \tag{4.6}$$

i.e., for the extended root, (4.5) is satisfied.

(2) Suppose $\sum_{i=1}^{n} 2^{-l_i} = 1$ holds for an arbitrary binary tree with n leaves. Now we extend one leaf, say the nth leaf.[4] We get a new tree with $n' = n + 1$ leaves, where

$$\sum_{i=1}^{n'} 2^{-l_i} = \underbrace{\sum_{i=1}^{n-1} 2^{-l_i}}_{\substack{\text{unchanged} \\ \text{leaves}}} + 2 \cdot \underbrace{2^{-(l_n+1)}}_{\substack{\text{new leaves} \\ \text{at depth} \\ l_n+1}} \tag{4.7}$$

$$= \sum_{i=1}^{n-1} 2^{-l_i} + 2^{-l_n} \tag{4.8}$$

$$= \sum_{i=1}^{n} 2^{-l_i} = 1. \tag{4.9}$$

Here the last equality follows from our assumption that $\sum_{i=1}^{n} 2^{-l_i} = 1$. Hence, by extending one leaf, the second statement continues to hold.

(3) Since any tree can be grown by continuously extending some leaves, the proof follows by induction. □

We are now ready to apply our first insights about trees to codes.

4.4 The Kraft Inequality

The following theorem is very useful because it gives us a way of finding out whether a prefix-free code exists or not.

[4] Since the tree is arbitrary, it does not matter how we number the leaves!

Theorem 4.8 (Kraft Inequality) *There exists a binary prefix-free code with r codewords of lengths $l_1, l_2, \ldots, l_r$ if, and only if,*

$$\sum_{i=1}^{r} 2^{-l_i} \leq 1. \tag{4.10}$$

If (4.10) is satisfied with equality, then there are no unused leaves in the tree.

Example 4.9 Let $l_1 = 3$, $l_2 = 4$, $l_3 = 4$, $l_4 = 4$, $l_5 = 4$. Then

$$2^{-3} + 4 \cdot 2^{-4} = \frac{1}{8} + \frac{4}{16} = \frac{3}{8} \leq 1; \tag{4.11}$$

i.e., there exists a binary prefix-free code consisting of five codewords with the given codeword lengths.

On the other hand, we cannot find any prefix-free code with five codewords of lengths $l_1 = 1$, $l_2 = 2$, $l_3 = 3$, $l_4 = 3$, and $l_5 = 4$ because

$$2^{-1} + 2^{-2} + 2 \cdot 2^{-3} + 2^{-4} = \frac{17}{16} > 1. \tag{4.12}$$

These two examples are shown graphically in Figure 4.6. ◊

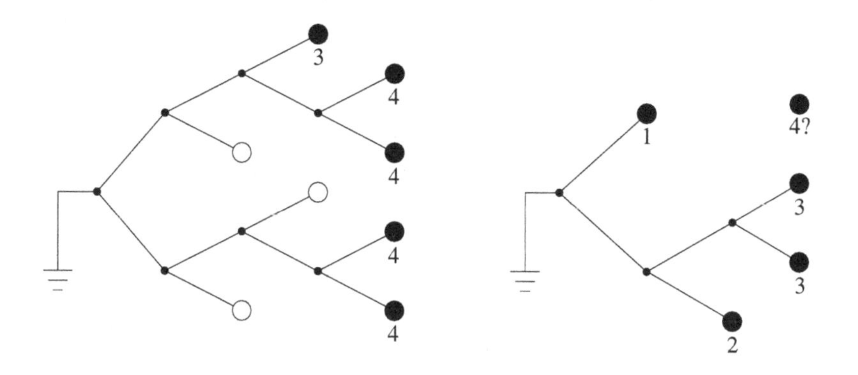

Figure 4.6 Examples of the Kraft Inequality.

Proof of the Kraft Inequality We prove the two directions separately.

$\Longrightarrow$: Suppose that there exists a binary prefix-free code with the given codeword lengths. From Lemma 4.4 we know that all r codewords of a binary prefix-free code are leaves in a binary tree. The total number n of (used and unused) leaves in this tree can therefore not be smaller than r, i.e.

$$r \leq n. \tag{4.13}$$

Hence,

$$\sum_{i=1}^{r} 2^{-l_i} \le \sum_{i=1}^{n} 2^{-l_i} = 1, \tag{4.14}$$

where the last equality follows from the Leaf-Depth Lemma (Lemma 4.7).

$\Longleftarrow$: Suppose that $\sum_{i=1}^{r} 2^{-l_i} \le 1$. We now can construct a prefix-free code as follows:

Step 1 Start with the extended root, i.e. a tree with two leaves, set $i = 1$, and assume, without loss of generality, that $l_1 \le l_2 \le \cdots \le l_r$.

Step 2 If there is an unused leaf at depth l_i, put the ith codeword there. Note that there could be none because l_i can be strictly larger than the current depth of the tree. In this case, extend any unused leaf to depth l_i, and put the ith codeword to one of the new leaves.

Step 3 If $i = r$, stop. Otherwise $i \to i + 1$ and go to Step 2.

We only need to check that Step 2 is always possible, i.e. that there is always some unused leaf available. To that goal, note that if we get to Step 2, we have already put $i - 1$ codewords into the tree. From the Leaf-Depth Lemma (Lemma 4.7) we know that

$$1 = \sum_{j=1}^{n} 2^{-\tilde{l}_j} = \underbrace{\sum_{j=1}^{i-1} 2^{-l_j}}_{\text{used leaves}} + \underbrace{\sum_{j=i}^{n} 2^{-\tilde{l}_j}}_{\text{unused leaves}}, \tag{4.15}$$

where $\tilde{l}_j$ are the depths of the leaves in the tree at that moment; i.e., $(\tilde{l}_1, \ldots, \tilde{l}_{i-1}) = (l_1, \ldots, l_{i-1})$ and $\tilde{l}_i, \ldots, \tilde{l}_n$ are the depths of the (so far) unused leaves. Now note that in our algorithm $i \le r$, i.e.

$$\sum_{j=1}^{i-1} 2^{-l_j} < \sum_{j=1}^{r} 2^{-l_j} \le 1, \tag{4.16}$$

where the last inequality follows by assumption. Hence,

$$\underbrace{\sum_{j=1}^{i-1} 2^{-l_j}}_{<1} + \sum_{j=i}^{n} 2^{-\tilde{l}_j} = 1 \quad \Longrightarrow \quad \sum_{j=i}^{n} 2^{-\tilde{l}_j} > 0 \tag{4.17}$$

and there still must be some unused leaves available! □

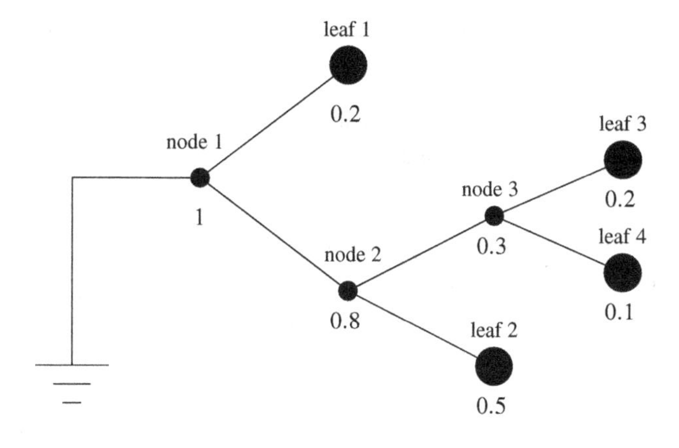

Figure 4.7 Rooted tree with probabilities.

4.5 Trees with probabilities

We have seen already in Section 4.1 that for codes it is important to consider the probabilities of the codewords. We therefore now introduce probabilities in our trees.

Definition 4.10 A *rooted tree with probabilities* is a finite rooted tree with probabilities assigned to each node and leaf such that

- the probability of a node is the sum of the probabilities of its children, and
- the root has probability 1.

An example of a rooted tree with probabilities is given in Figure 4.7. Note that the probabilities can be seen as the overall probability of passing through a particular node (or reaching a particular leaf) when making a random walk from the root to a leaf. Since we start at the root, the probability that our path goes through the root is always 1. Then, in the example of Figure 4.7, we have an 80% chance that our path will go through node 2 and a 10% chance to end up in leaf 4.

Since in a prefix-free code all codewords are leaves and we are particularly interested in the average codeword length, we are very much interested in the average depth of the leaves in a tree (where for the averaging operation we use the probabilities in the tree). Luckily, there is an elegant way to compute this average depth, as shown in the following lemma.

Lemma 4.11 (Path Length Lemma) *In a rooted tree with probabilities, the* *average depth* L_{av} *of the leaves is equal to the sum of the probabilities of all nodes (including the root).*

To clarify our notation we refer to leaf probabilities by small p_i while node probabilities are denoted by capital P_ℓ.

Example 4.12 Consider the tree of Figure 4.7. We have four leaves: one at depth $l_1 = 1$ with a probability $p_1 = 0.2$, one at depth $l_2 = 2$ with a probability $p_2 = 0.5$, and two at depth $l_3 = l_4 = 3$ with probabilities $p_3 = 0.2$ and $p_4 = 0.1$, respectively. Hence, the average depth of the leaves is given by

$$\mathsf{L}_{\text{av}} = \sum_{i=1}^{4} p_i l_i = 0.2 \cdot 1 + 0.5 \cdot 2 + 0.2 \cdot 3 + 0.1 \cdot 3 = 2.1. \tag{4.18}$$

According to Lemma 4.11, this must be equal to the sum of the node probabilities:

$$\mathsf{L}_{\text{av}} = P_1 + P_2 + P_3 = 1 + 0.8 + 0.3 = 2.1. \tag{4.19}$$

◊

Proof of Lemma 4.11 The lemma is easiest understood when looking at a particular example. Let us again consider the tree of Figure 4.7: the probability $p_1 = 0.2$ of leaf 1 needs to be counted once only, which is the case as it is only part of the probability of the root $P_1 = 1$. The probability $p_2 = 0.5$ must be counted twice. This is also the case because it is contained in the root probability $P_1 = 1$ and also in the probability of the second node $P_2 = 0.8$. Finally, the probabilities of leaf 3 and leaf 4, $p_3 = 0.2$ and $p_4 = 0.1$, are counted three times: they are part of P_1, P_2, and P_3:

$$\begin{aligned}
\mathsf{L}_{\text{av}} &= 2.1 && (4.20)\\
&= 1 \cdot 0.2 + 2 \cdot 0.5 + 3 \cdot 0.2 + 3 \cdot 0.1 && (4.21)\\
&= 1 \cdot (0.2 + 0.5 + 0.2 + 0.1) + 1 \cdot (0.5 + 0.2 + 0.1) \\
&\quad + 1 \cdot (0.2 + 0.1) && (4.22)\\
&= 1 \cdot P_1 + 1 \cdot P_2 + 1 \cdot P_3 && (4.23)\\
&= P_1 + P_2 + P_3. && (4.24)
\end{aligned}$$

□

4.6 Optimal codes: Huffman code

Let us now connect the probabilities in the tree with the probabilities of the code, or actually, more precisely, the probabilities of the random messages that shall be represented by the code. We assume that we have in total r different message symbols. Let the probability of the ith message symbol be p_i and let the length of the corresponding codeword representing this symbol be l_i. Then the average length of the code is given by

$$\mathsf{L}_{\text{av}} = \sum_{i=1}^{r} p_i l_i. \tag{4.25}$$

With no loss in generality, the p_i may be taken in nonincreasing order. If the lengths l_i are then not in the opposite order, i.e. we do not have both

$$p_1 \geq p_2 \geq p_3 \geq \cdots \geq p_r \tag{4.26}$$

and

$$l_1 \leq l_2 \leq l_3 \leq \cdots \leq l_r, \tag{4.27}$$

then the code is not *optimal* in the sense that we could have a shorter average length by reassigning the codewords to different symbols. To prove this claim, suppose that for some i and j with $i < j$ we have both

$$p_i > p_j \qquad \text{and} \qquad l_i > l_j. \tag{4.28}$$

In computing the average length, originally, the sum in (4.25) contains, among others, the two terms

$$\text{old:} \qquad p_i l_i + p_j l_j. \tag{4.29}$$

By interchanging the codewords for the ith and jth symbols, we get the terms

$$\text{new:} \qquad p_i l_j + p_j l_i, \tag{4.30}$$

while the remaining terms are unchanged. Subtracting the old from the new we see that

$$\begin{aligned} \text{new} - \text{old:} \qquad (p_i l_j + p_j l_i) - (p_i l_i + p_j l_j) &= p_i(l_j - l_i) + p_j(l_i - l_j) && (4.31) \\ &= (p_i - p_j)(l_j - l_i) && (4.32) \\ &< 0. && (4.33) \end{aligned}$$

From (4.28) this is a negative number, i.e. we can decrease the average codeword length by interchanging the codewords for the ith and jth symbols. Hence the new code with exchanged codewords for the ith and jth symbols is better than the original code – which therefore cannot have been optimal.

We will now examine the optimal binary code which is called the *Huffman code* due to its discoverer. The trick of the derivation of the optimal code is the insight that the corresponding code tree has to be grown backwards, starting

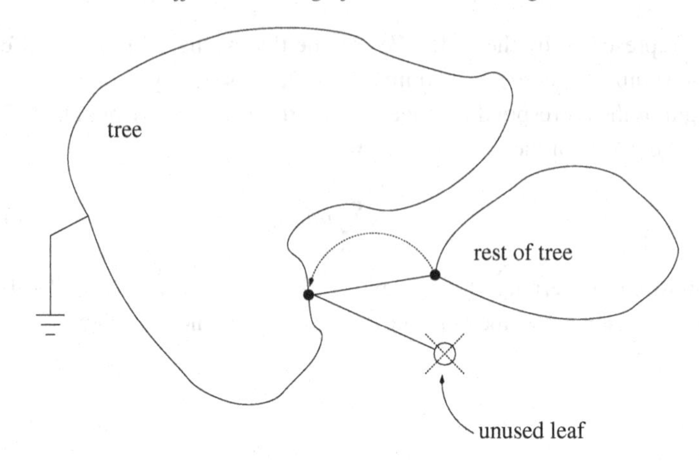

Figure 4.8 Code performance and unused leaves: by deleting the unused leaf and moving its sibling to the parent, we can improve on the code's performance.

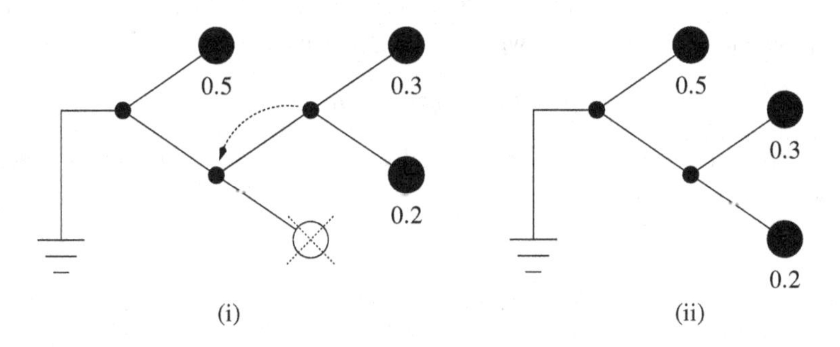

Figure 4.9 Improving a code by removing an unused leaf.

from the leaves (and not, as might be intuitive at a first glance, starting from the root).

The clue of binary Huffman coding lies in two basic observations. The first observation is as follows.

Lemma 4.13 *In a binary tree of an optimal binary prefix-free code, there is no unused leaf.*

Proof Suppose that the tree of an optimal code has an unused leaf. Then we can delete this leaf and move its sibling to the parent node; see Figure 4.8. By doing so we reduce L_{av}, which contradicts our assumption that the original code was optimal. □

Example 4.14 As an example consider the two codes given in Figure 4.9, both of which have three codewords. Code (i) has an average length[5] of $\mathsf{L}_{\mathrm{av}} = 2$, and code (ii) has an average length of $\mathsf{L}_{\mathrm{av}} = 1.5$. Obviously, code (ii) performs better. ◊

The second observation basically says that the two most unlikely symbols must have the longest codewords.

Lemma 4.15 *There exists an optimal binary prefix-free code such that the two least likely codewords only differ in the last digit, i.e. the two most unlikely codewords are siblings.*

Proof Since we consider an optimal code, the codewords that correspond to the two least likely symbols must be the longest codewords (see our discussion after (4.27)). If they have the same parent node, we are done. If they do not have the same parent node, this means that there exist other codewords of the same length (because we know from Lemma 4.13 that there are no unused leaves). In this case, we can simply swap two codewords of equal maximum length in such a way that the two least likely codewords have the same parent, and we are done. □

Because of Lemma 4.13 and the Path Length Lemma (Lemma 4.11), we see that the construction of an optimal binary prefix-free code for an r-ary random message U is equivalent to constructing a binary tree with r leaves such that the sum of the probabilities of the nodes is minimum when the leaves are assigned the probabilities p_i for $i = 1, 2, \ldots, r$:

$$\mathsf{L}_{\mathrm{av}} = \underbrace{\sum_{\ell=1}^{\mathsf{N}} P_\ell}_{\longrightarrow \text{minimize!}} . \tag{4.34}$$

But Lemma 4.15 tells us how we may choose one node in an optimal code tree, namely as the parent of the two least likely leaves p_{r-1} and p_r:

$$P_{\mathsf{N}} = p_{r-1} + p_r. \tag{4.35}$$

So we have fixed one P_ℓ in (4.34) already. But, if we now pruned our binary tree at this node to make it a leaf with probability $p = p_{r-1} + p_r$, it would become one of $(r-1)$ leaves in a new tree. Completing the construction of the optimal code would then be equivalent to constructing a binary tree with these

[5] Remember Lemma 4.11 to compute the average codeword length: summing the node probabilities. In code (i) we have $P_1 = 1$ and $P_2 = P_3 = 0.5$ (note that the unused leaf has, by definition, zero probability), and in code (ii) $P_1 = 1$ and $P_2 = 0.5$.

$(r-1)$ leaves such that the sum of the probabilities of the nodes is minimum:

$$\mathrm{L}_{\mathrm{av}} = \underbrace{\sum_{\ell=1}^{\mathrm{N}-1} P_\ell}_{\longrightarrow \text{minimize!}} + \underbrace{P_{\mathrm{N}}}_{\substack{\text{optimally}\\ \text{chosen}}}. \tag{4.36}$$

Again Lemma 4.15 tells us how to choose one node in this new tree, and so on. We have thus proven the validity of the following algorithm.

> **Huffman's Algorithm for Optimal Binary Codes**
>
> **Step 1** Create r leaves corresponding to the r possible symbols and assign their probabilities $p_1, \ldots, p_r$. Mark these leaves as active.
>
> **Step 2** Create a new node that has the two *least likely* active leaves or nodes as children. Activate this new node and deactivate its children.
>
> **Step 3** If there is only one active node left, root it. Otherwise, go to Step 2.

Example 4.16 In Figure 4.10 we show the procedure of producing a Huffman code for the example of a random message with four possible symbols with probabilities $p_1 = 0.4$, $p_2 = 0.3$, $p_3 = 0.2$, $p_4 = 0.1$. We see that the average codeword length of this Huffman code is

$$\mathrm{L}_{\mathrm{av}} = 0.4 \cdot 1 + 0.3 \cdot 2 + 0.2 \cdot 3 + 0.1 \cdot 3 = 1.9. \tag{4.37}$$

Using Lemma 4.11 this can be computed much easier as follows:

$$\mathrm{L}_{\mathrm{av}} = P_1 + P_2 + P_3 = 1 + 0.6 + 0.3 = 1.9. \tag{4.38}$$

◊

Note that the code design process is not unique in several respects. Firstly, the assignment of the 0 or 1 digits to the codewords at each forking stage is arbitrary, but this produces only trivial differences. Usually, we will stick to the convention that going upwards corresponds to 0 and downwards to 1. Secondly, when there are more than two least likely (active) nodes/leaves, it does not matter which we choose to combine. The resulting codes can have codewords of different lengths; however, the average codeword length will always be the same.

Example 4.17 As an example of different Huffman encodings of the same random message, let $p_1 = 0.4$, $p_2 = 0.2$, $p_3 = 0.2$, $p_4 = 0.1$, $p_5 = 0.1$. Figure 4.11 shows three different Huffman codes for this message: the list of

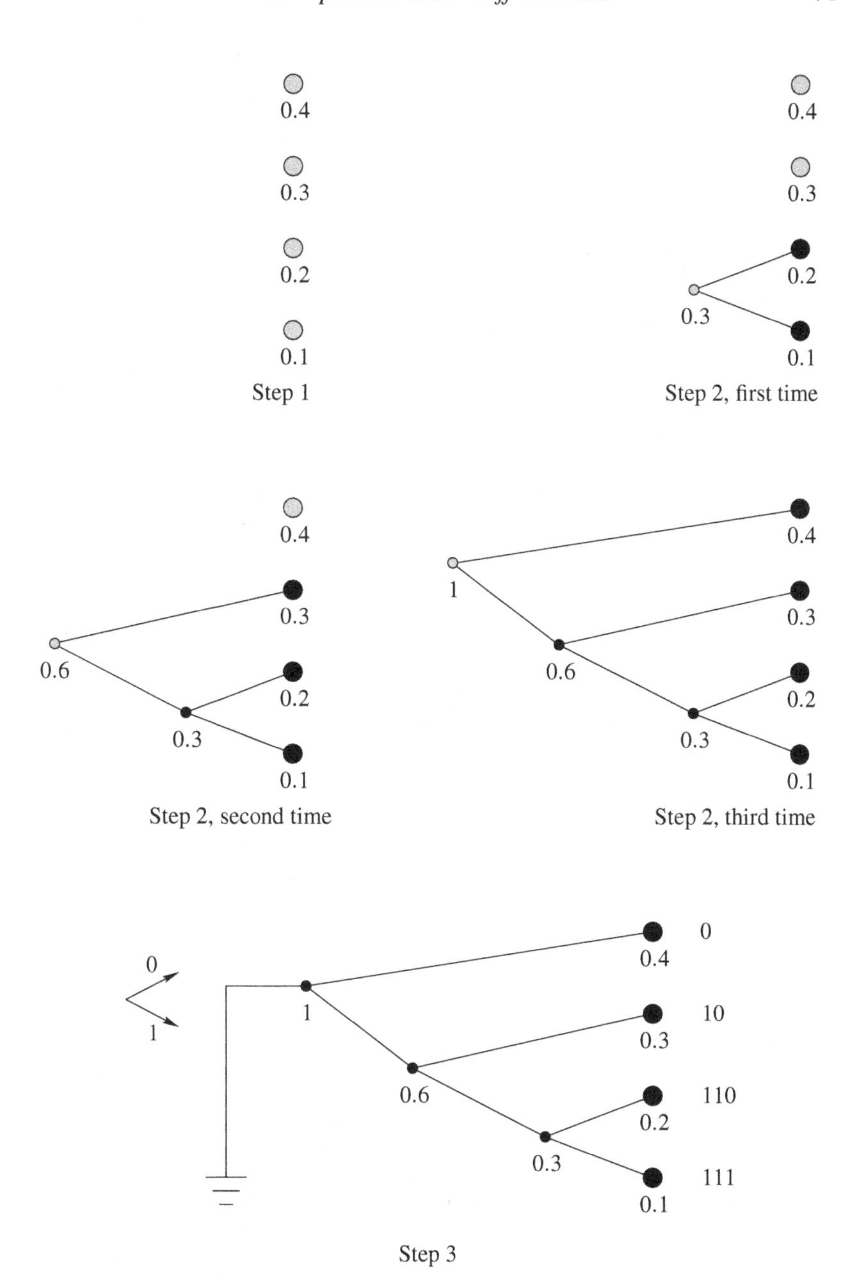

Figure 4.10 Creation of a binary Huffman code. Active nodes and leaves are shaded.

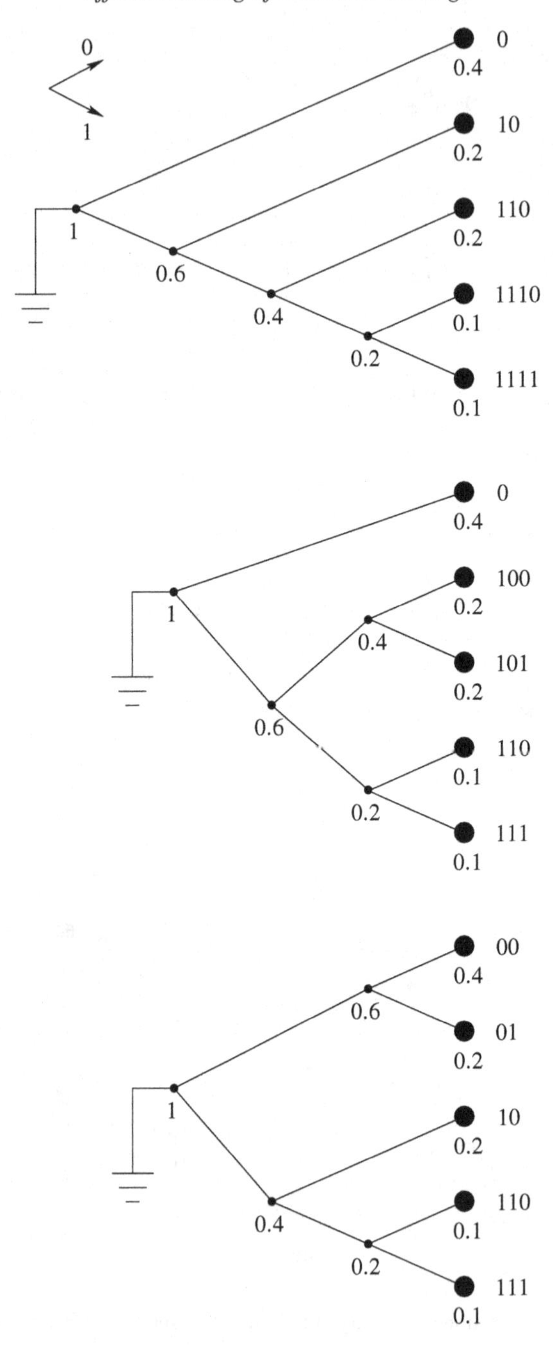

Figure 4.11 Different binary Huffman codes for the same random message.

codeword lengths are $(1,2,3,4,4)$, $(1,3,3,3,3)$, and $(2,2,2,3,3)$, respectively. But all of these codes have the same performance, $\mathsf{L}_{\text{av}} = 2.2$. ◊

Exercise 4.18 *Try to generate all three codes of Example 4.17 (see Figure 4.11) yourself.* ◊

4.7 Types of codes

Note that in Section 4.1 we have restricted ourselves to prefix-free codes. So, up to now we have only proven that Huffman codes are the optimal codes *under the assumption that we restrict ourselves to prefix-free codes.* We would now like to show that Huffman codes are actually optimal among all useful codes.

To reach that goal, we need to come back to a more precise definition of "useful codes," i.e. we continue the discussion that we started in Section 4.1. Let us consider an example with a random message U with four different symbols and let us design various codes for this message as shown in Table 4.2.

Table 4.2 *Various codes for a random message with four possible values*

U	Code (i)	Code (ii)	Code (iii)	Code (iv)
a	0	0	10	0
b	0	010	00	10
c	1	01	11	110
d	1	10	110	111

We discuss these different codes.

Code (i) is useless because some codewords are used for more than one symbol. Such a code is called *singular.*

Code (ii) is nonsingular. But we have another problem: if we receive 010 we have three different possibilities how to decode it: it could be (010) giving us b, or it could be (0)(10) leading to ad, or it could be (01)(0) corresponding to ca. Even though nonsingular, this code is not *uniquely decodable* and therefore in practice is as useless as code (i).[6]

[6] Note that adding a comma between the codewords is not allowed because in this case we change the code to be *ternary*, i.e. the codewords contain three different letters "0", "1", and "," instead of only two "0" and "1". By the way, it is not very difficult to generalize all results given in this chapter to D-ary codes. See, for example, [Mas96]. In this book, we will stick to binary codes.

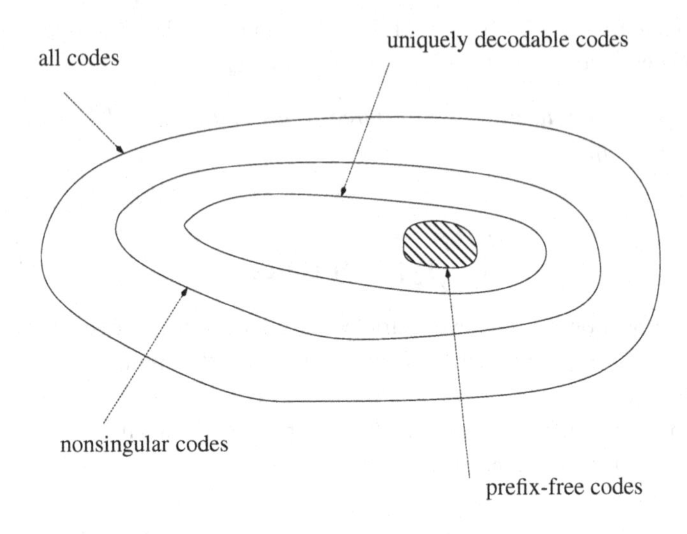

Figure 4.12 Set of all codes.

Code (iii) is uniquely decodable, even though it is not prefix-free! To see this, note that in order to distinguish between c and d we only need to wait for the next 1 to show up: if the number of 0s in between is even, we decode 11, otherwise we decode 110. Example:

$$11000010 = (11)(00)(00)(10) \implies cbba, \tag{4.39}$$

$$110000010 = (110)(00)(00)(10) \implies dbba. \tag{4.40}$$

So in a uniquely decodable but not prefix-free code we may have to delay the decoding until later.

Code (iv) is prefix-free and therefore trivially uniquely decodable.

We see that the set of all possible codes can be grouped as shown in Figure 4.12. We are only interested in the uniquely decodable codes. But so far we have restricted ourselves to prefix-free codes. So the following question arises: is there a uniquely decodable code that is not prefix-free, but that has a better performance than the best prefix-free code (i.e. the corresponding Huffman code)?

Luckily the answer to this question is No, i.e. the Huffman codes are the best uniquely decodable codes. This can be seen from the following theorem.

Theorem 4.19 (McMillan's Theorem) *The codeword lengths l_i of any* *uniquely decodable code must satisfy the Kraft Inequality*

$$\sum_{i=1}^{r} 2^{-l_i} \le 1. \tag{4.41}$$

Why does this help to answer our question about the most efficient uniquely decodable code? Well, note that we know from Theorem 4.8 that every prefix-free code also satisfies (4.41). So, for any uniquely decodable, but non-prefix-free code with given codeword lengths, one can find another code with the same codeword lengths that is prefix-free. But if the codeword lengths are the same, the performance is identical! Hence, there is no gain in designing a non-prefix-free code.

Proof of Theorem 4.19 Suppose we are given a random message U that takes on r possible values $u \in \mathcal{U}$ (here the set $\mathcal{U}$ denotes the message alphabet). Suppose further that we have a *uniquely decodable* code that assigns to every possible symbol $u \in \mathcal{U}$ a certain codeword of length $l(u)$.

Now choose an arbitrary positive integer ν and design a new code for a vector of ν symbols $\mathbf{u} = (u_1, u_2, \ldots, u_\nu) \in \mathcal{U}^\nu = \mathcal{U} \times \cdots \times \mathcal{U}$ by simply *concatenating* the original codewords.

Example 4.20 Consider a ternary message with the possible values $u = a$, $u = b$, or $u = c$, i.e. $\mathcal{U} = \{a, b, c\}$. If the probabilities of these possible values are

$$\Pr[U = a] = \frac{1}{2}, \quad \Pr[U = b] = \Pr[U = c] = \frac{1}{4}, \tag{4.42}$$

a binary (single-letter) Huffman code would map

$$a \mapsto 0, \qquad b \mapsto 10, \qquad c \mapsto 11. \tag{4.43}$$

If we now choose $\nu = 3$, we get a new source with $3^3 = 27$ possible symbols, namely

$$\begin{aligned} \mathcal{U}^3 = \{&aaa, aab, aac, aba, abb, abc, aca, acb, acc, \\ &baa, bab, bac, bba, bbb, bbc, bca, bcb, bcc, \\ &caa, cab, cac, cba, cbb, cbc, cca, ccb, ccc\}. \end{aligned} \tag{4.44}$$

The corresponding 27 codewords are then as follows (given in the same order):

$$\begin{aligned} \{&000, 0010, 0011, 0100, 01010, 01011, 0110, 01110, 01111, \\ &1000, 10010, 10011, 10100, 101010, 101011, 10110, 101110, 101111, \\ &1100, 11010, 11011, 11100, 111010, 111011, 11110, 111110, 111111\}. \end{aligned} \tag{4.45}$$

The clue observation now is that because the original code was uniquely decodable, it immediately follows that this new concatenated code also must be uniquely decodable.

Exercise 4.21 *Explain this clue observation, i.e. explain why the new concatenated code is also uniquely decodable.*

Hint: Note that the codewords of the new code consist of a sequence of uniquely decodable codewords. ♢

The lengths of the new codewords are given by

$$\tilde{l}(\mathbf{u}) = \sum_{j=1}^{\nu} l(u_j). \tag{4.46}$$

Let $l_{\max}$ be the maximal codeword length of the original code. Then the new code has a maximal codeword length $\tilde{l}_{\max}$ satisfying

$$\tilde{l}_{\max} = \nu l_{\max}. \tag{4.47}$$

We now compute the following:

$$\left(\sum_{u\in\mathcal{U}} 2^{-l(u)}\right)^{\nu} = \left(\sum_{u_1\in\mathcal{U}} 2^{-l(u_1)}\right)\left(\sum_{u_2\in\mathcal{U}} 2^{-l(u_2)}\right)\cdots\left(\sum_{u_\nu\in\mathcal{U}} 2^{-l(u_\nu)}\right) \tag{4.48}$$

$$= \sum_{u_1\in\mathcal{U}}\sum_{u_2\in\mathcal{U}}\cdots\sum_{u_\nu\in\mathcal{U}} 2^{-l(u_1)}2^{-l(u_2)}\cdots 2^{-l(u_\nu)} \tag{4.49}$$

$$= \sum_{\mathbf{u}\in\mathcal{U}^\nu} 2^{-l(u_1)-l(u_2)-\cdots-l(u_\nu)} \tag{4.50}$$

$$= \sum_{\mathbf{u}\in\mathcal{U}^\nu} 2^{-\sum_{j=1}^{\nu} l(u_j)} \tag{4.51}$$

$$= \sum_{\mathbf{u}\in\mathcal{U}^\nu} 2^{-\tilde{l}(\mathbf{u})}. \tag{4.52}$$

Here (4.48) follows by writing the exponentiated sum as a product of ν sums; in (4.50) we combine the ν sums over $u_1,\ldots,u_\nu$ into one huge sum over the ν-vector $\mathbf{u}$; and (4.52) follows from (4.46).

Next we will rearrange the order of the terms by collecting all terms with the same exponent together:

$$\sum_{\mathbf{u}\in\mathcal{U}^\nu} 2^{-\tilde{l}(\mathbf{u})} = \sum_{m=1}^{\tilde{l}_{\max}} w(m)2^{-m}, \tag{4.53}$$

where $w(m)$ counts the number of such terms with equal exponent, i.e. $w(m)$ denotes the number of codewords of length m in the new code.

Example 4.22 (Continuation from Example 4.20) We see from (4.45) that the new concatenated code has one codeword of length 3, six codewords of length 4, twelve codewords of length 5, and eight codewords of length 6. Hence,

$$\sum_{\mathbf{u}\in\mathcal{U}^{\nu}} 2^{-\tilde{l}(\mathbf{u})} = 1\cdot 2^{-3} + 6\cdot 2^{-4} + 12\cdot 2^{-5} + 8\cdot 2^{-6}, \tag{4.54}$$

i.e.

$$w(m) = \begin{cases} 1 & \text{for } m = 3, \\ 6 & \text{for } m = 4, \\ 12 & \text{for } m = 5, \\ 8 & \text{for } m = 6, \\ 0 & \text{otherwise.} \end{cases} \tag{4.55}$$

Also note that $\tilde{l}_{\max} = 6 = \nu \cdot l_{\max} = 3 \cdot 2$ in this case. ◊

We combine (4.53) and (4.52) and use (4.47) to write

$$\left(\sum_{u\in\mathcal{U}} 2^{-l(u)}\right)^{\nu} = \sum_{m=1}^{\nu l_{\max}} w(m)2^{-m}. \tag{4.56}$$

Note that since the new concatenated code is uniquely decodable, every codeword of length m is used at most once. But in total there are only 2^m different sequences of length m, i.e. we know that

$$w(m) \le 2^m. \tag{4.57}$$

Thus,

$$\left(\sum_{u\in\mathcal{U}} 2^{-l(u)}\right)^{\nu} = \sum_{m=1}^{\nu l_{\max}} w(m)2^{-m} \le \sum_{m=1}^{\nu l_{\max}} 2^m 2^{-m} = \nu l_{\max} \tag{4.58}$$

or

$$\sum_{u\in\mathcal{U}} 2^{-l(u)} \le (\nu l_{\max})^{1/\nu}. \tag{4.59}$$

At this stage we are back to an expression that depends only on the original uniquely decodable code. So forget about the trick with the new concatenated code, but simply note that we have shown that for any uniquely decodable code and any positive integer ν, expression (4.59) must hold! Also note that we can choose ν freely here.

Note further that for any finite value of $l_{\max}$ one can show that

$$\lim_{\nu\to\infty} (\nu l_{\max})^{1/\nu} = 1. \tag{4.60}$$

Hence, by choosing ν extremely large (i.e. we let ν tend to infinity) we have

$$\sum_{u \in \mathcal{U}} 2^{-l(u)} \leq 1 \tag{4.61}$$

as we wanted to prove. □

4.8 Some historical background

David A. Huffman had finished his B.S. and M.S. in electrical engineering and also served in the U.S. Navy before he became a Ph.D. student at the Massachusetts Institute of Technology (MIT). There, in 1951, he attended an information theory class taught by Professor Robert M. Fano who was working at that time, together with Claude E. Shannon, on finding the most efficient code, but could not solve the problem. So Fano assigned the question to his students in the information theory class as a term paper. Huffman tried for a long time to find a solution and was about to give up when he had the sudden inspiration to start building the tree backwards from leaves to root instead from root to leaves. Once he had understood this, he was quickly able to prove that his code was the most efficient one. Naturally, Huffman's term paper was later published.

Huffman became a faculty member of MIT in 1953, and later, in 1967, he moved to the University of California, Santa Cruz, where he stayed until his retirement in 1994. He won many awards for his accomplishments, e.g. in 1988 the Richard Hamming Medal from the Institute of Electrical and Electronics Engineers (IEEE). Huffman died in 1998. See [Sti91] and [Nor89].

4.9 Further reading

For an easy-to-read, but precise, introduction to coding and trees, the lecture notes [Mas96] of Professor James L. Massey from ETH Zurich are highly recommended. There the interested reader will find a straightforward way to generalize the concept of binary codes to general D-ary codes using D-ary trees. Moreover, in [Mas96] one also finds the concept of *block codes*, i.e. codes with a fixed codeword length. Some of the best such codes are called *Tunstall codes* [Tun67].

References

[Mas96] James L. Massey, *Applied Digital Information Theory I and II,* Lecture notes, Signal and Information Processing Laboratory, ETH Zurich, 1995/1996. Available: http://www.isiweb.ee.ethz.ch/archive/massey_scr/

[Nor89] Arthur L. Norberg, "An interview with Robert M. Fano," Charles Babbage Institute, Center for the History of Information Processing, April 1989.

[Sti91] Gary Stix, "Profile: Information theorist David A. Huffman," *Scientific American (Special Issue on Communications, Computers, and Networks)*, vol. 265, no. 3, September 1991.

[Tun67] Brian P. Tunstall, "Synthesis of noiseless compression codes," Ph.D. dissertation, Georgia Institute of Technology, September 1967.

5

Entropy and Shannon's Source Coding Theorem

Up to this point we have been concerned with coding theory. We have described codes and given algorithms of how to design them. And we have evaluated the performance of some *particular* codes. Now we begin with information theory, which will enable us to learn more about the fundamental properties of *general* codes without having actually to design them.

Basically, information theory is a part of physics and tries to describe what information is and how we can work with it. Like all theories in physics it is a model of the real world that is accepted as true as long as it predicts how nature behaves accurately enough.

In the following we will start by giving some suggestive examples to motivate the definitions that follow. However, note that these examples are not a justification for the definitions; they just try to shed some light on the reason why we will define these quantities in the way we do. The real justification of all definitions in information theory (or any other physical theory) is the fact that they turn out to be useful.

5.1 Motivation

We start by asking the question: what is information?

Let us consider some examples of sentences that contain some "information."

- The weather will be good tomorrow.
- The weather was bad last Sunday.
- The president of Taiwan will come to you tomorrow and will give you one million dollars.

The second statement seems not very interesting as you might already know what the weather was like last Sunday. The last statement is much more exciting than the first two and therefore seems to contain much more information. But, on the other hand, do you actually believe it? Do you think it is likely that you will receive one million dollars tomorrow?

Let us consider some easier examples.

- You ask: "Is the temperature in Taiwan currently above 30 degrees?"

  This question has only two possible answers: "yes" or "no."
- You ask: "The president of Taiwan has spoken with a certain person from Hsinchu today. With whom?"

  Here, the question has about 400 000 possible answers (since Hsinchu has about 400 000 inhabitants).

Obviously the second answer provides you with a much bigger amount of information than the first one. We learn the following.

> The number of possible answers r should be linked to "information."

Here is another example.

- You observe a gambler throwing a fair dice. There are six possible outcomes $\{1,2,3,4,5,6\}$. You note the outcome and then tell it to a friend. By doing so you give your friend a certain amount of information.
- Next you observe the gambler throwing the dice *three times*. Again, you note the three outcomes and tell them to your friend. Obviously, the amount of information that you give to your friend this time is three times as much as the first time.

So we learn the following.

> "Information" should be additive in some sense.

Now we face a new problem: regarding the example of the gambler above we see that in the first case we have $r = 6$ possible answers, while in the second case we have $r = 6^3 = 216$ possible answers. Hence in the second experiment there are 36 times more possible outcomes than in the first experiment! But we would like to have only a three times larger amount of information. So how do we solve this?

Idea: use a logarithm. Then the exponent 3 will become a factor exactly as we wish: $\log_b 6^3 = 3 \cdot \log_b 6$.

Exactly these observations have been made by the researcher Ralph Hartley in 1928 in Bell Labs [Har28]. He gave the following definition.

Definition 5.1 We define the following measure of information:

$$\tilde{I}(U) \triangleq \log_b r, \tag{5.1}$$

where r is the number of all possible outcomes of a random message U.

Using this definition we can confirm that it has the wanted property of additivity:

$$\tilde{I}(U_1, U_2, \ldots, U_n) = \log_b r^n = n \cdot \log_b r = n \cdot \tilde{I}(U). \tag{5.2}$$

Hartley also correctly noted that the basis b of the logarithm is not really important for this measure. It only decides on the *unit of information*. So, similarly to the fact that 1 km is the same distance as 1000 m, b is only a change of units without actually changing the amount of information it describes.

For two important and one unimportant special cases of b it has been agreed to use the following names for these units:

$$\begin{aligned} &b = 2\ (\log_2): &&\text{bit,}\\ &b = e\ (\ln): &&\text{nat (natural logarithm),}\\ &b = 10\ (\log_{10}): &&\text{Hartley.} \end{aligned}$$

Note that the unit Hartley has been chosen in honor of the first researcher who made the first (partially correct) attempt at defining information. However, as nobody in the world ever uses the basis $b = 10$ for measuring information, this honor is questionable.

The measure $\tilde{I}(U)$ is the right answer to many technical problems.

Example 5.2 A village has eight telephones. How long must the phone number be? Or, asked differently: how many bits of information do we need to send to the central office so that we are connected to a particular phone?

$$8 \text{ phones} \quad \Longrightarrow \quad \log_2 8 = 3 \text{ bits}. \tag{5.3}$$

We choose the following phone numbers:

$$\{000, 001, 010, 011, 100, 101, 110, 111\}. \tag{5.4}$$

$\diamondsuit$

In spite of its usefulness, Hartley's definition had no effect whatsoever in the world. That's life... On the other hand, it must be admitted that Hartley's definition has a fundamental flaw. To realize that something must be wrong, note that according to (5.1) the *smallest* nonzero amount of information is

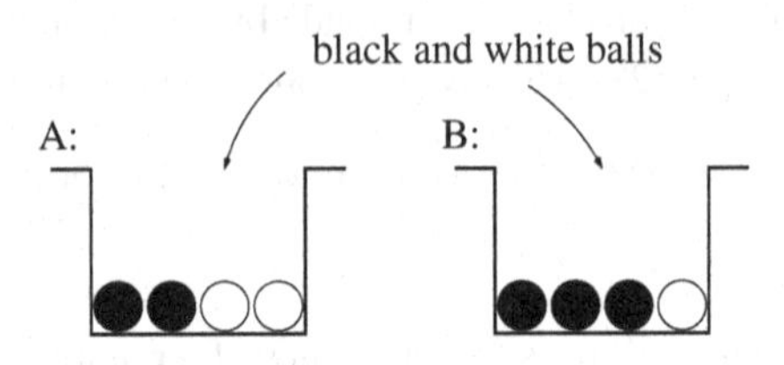

Figure 5.1 Two hats with four balls each.

$\log_2 2 = 1$ bit. This might sound like only a small amount of information, but actually 1 bit can be *a lot* of information! As an example, consider the 1-bit (yes or no) answer if a man asks a woman whether she wants to marry him. If you still do not believe that one bit is a huge amount of information, consider the following example.

Example 5.3 Currently there are 6 902 106 897 persons living on our planet (U.S. Census Bureau, 25 February 2011, 13:43 Taiwan time). How long must a binary telephone number U be if we want to be able to connect to every person?

According to Hartley we need $\tilde{I}(U) = \log_2(6902106897) \simeq 32.7$ bits. So with only 33 bits we can address every single person on this planet. Or, in other words, we only need 33 times 1 bit in order to distinguish every human being alive. $\diamond$

We see that 1 bit is a lot of information and it cannot be that this is the smallest amount of (nonzero) information.

To understand more deeply what is wrong, consider the two hats shown in Figure 5.1. Each hat contains four balls, where the balls can be either white or black. Let us draw one ball at random and let U be the color of the ball. In hat A we have $r = 2$ colors: black and white, i.e. $\tilde{I}(U_A) = \log_2 2 = 1$ bit. In hat B we also have $r = 2$ colors and hence also $\tilde{I}(U_B) = 1$ bit. But obviously, we get less information if in hat B black shows up, since we somehow expect black to show up in the first place. Black is much more *likely*!

We realize the following.

> A proper measure of information needs to take into account the probabilities of the various possible events.

This was observed for the first time by Claude Elwood Shannon in 1948 in his landmark paper "A mathematical theory of communication" [Sha48]. This paper has been like an explosion in the research community![1]

Before 1948, the engineering community was mainly interested in the behavior of a sinusoidal waveform that is passed through a communication system. Shannon, however, asked why we want to transmit a deterministic sinusoidal signal. The receiver already knows in advance that it will be a sinus, so it is much simpler to generate one at the receiver directly rather than to transmit it over a channel! In other words, Shannon had the fundamental insight that we need to consider *random* messages rather than deterministic messages whenever we deal with information.

Let us go back to the example of the hats in Figure 5.1 and have a closer look at hat B.

- There is one chance out of four possibilities that we draw a white ball.

  Since we would like to use Hartley's measure here, we recall that the quantity r inside the logarithm in (5.1) is "the number of all possible outcomes of a random message." Hence, from Hartley's point of view, we will see one realization out of r possible realizations. Translated to the case of the white ball, we see that we have one realization out of four possible realizations, i.e.

$$\log_2 4 = 2 \text{ bits} \tag{5.5}$$

  of information.

- On the other hand, there are three chances out of four that we draw a black ball.

  Here we cannot use Hartley's measure directly. But it is possible to translate the problem into a form that makes it somehow accessible to Hartley: we need to "normalize" the statement into a form that gives us *one* realization out of r. This can be done if we divide everything by 3, the number of black balls: we have one chance out of $4/3$ possibilities (whatever this means), or, stated differently, we have one realization out of $4/3$ possible "realizations," i.e.

$$\log_2 \frac{4}{3} = 0.415 \text{ bits} \tag{5.6}$$

  of information.

[1] Besides the amazing accomplishment of inventing information theory, at the age of 21 Shannon also "invented" the computer in his *Master* thesis [Sha37]! He proved that electrical circuits can be used to perform logical and mathematical operations, which was the foundation of digital computer and digital circuit theory. It is probably the most important Master thesis of the twentieth century! Incredible, isn't it?

So now we have two different values depending on what color we get. How shall we combine them to one value that represents the information? The most obvious choice is to average it, i.e. we weight the different information values according to their probabilities of occurrence:

$$\frac{1}{4} \cdot 2 \text{ bits} + \frac{3}{4} \cdot 0.415 \text{ bits} = 0.811 \text{ bits} \tag{5.7}$$

or

$$\frac{1}{4} \log_2 4 + \frac{3}{4} \log_2 \frac{4}{3} = 0.811 \text{ bits.} \tag{5.8}$$

We see the following.

> Shannon's measure of information is an "average Hartley information":
>
> $$\sum_{i=1}^{r} p_i \log_2 \frac{1}{p_i} = -\sum_{i=1}^{r} p_i \log_2 p_i, \tag{5.9}$$
>
> where p_i denotes the probability of the ith possible outcome.

We end this introductory section by pointing out that the given three motivating ideas, i.e.

(1) the number of possible answers r should be linked to "information";
(2) "information" should be additive in some sense; and
(3) a proper measure of information needs to take into account the probabilities of the various possible events,

are not sufficient to exclusively specify (5.9). The interested reader can find in Appendix 5.8 some more information on why Shannon's measure should be defined like (5.9) and not differently.

5.2 Uncertainty or entropy

5.2.1 Definition

We now formally define the Shannon measure of "self-information of a source." Due to its relationship with a corresponding concept in different areas of physics, Shannon called his measure *entropy*. We will stick to this name as it is standard in the whole literature. However, note that *uncertainty* would be a far more precise description.

Definition 5.4 (Entropy) The *uncertainty* or *entropy* of a random message U that takes on r different values with probability p_i, $i = 1,\ldots,r$, is defined as

$$H(U) \triangleq -\sum_{i=1}^{r} p_i \log_b p_i. \tag{5.10}$$

Remark 5.5 What happens if $p_i = 0$? Remember that $\log_b 0 = -\infty$. However, also note that $p_i = 0$ means that the symbol i never shows up. It therefore should not contribute to the uncertainty. Luckily this is the case:

$$\lim_{t \to 0} t \log_b t = 0, \tag{5.11}$$

i.e. we do not need to worry about this case.

So we note the following.

> Whenever we sum over $p_i \log_b p_i$, we implicitly assume that we exclude all indices i with $p_i = 0$.

As in the case of the Hartley measure of information, b denotes the *unit* of uncertainty:

$$b = 2: \quad \text{bit,} \tag{5.12}$$

$$b = e: \quad \text{nat,} \tag{5.13}$$

$$b = 10: \quad \text{Hartley.} \tag{5.14}$$

If the base of the logarithm is not specified, then we can choose it freely. However, note that the units are very important. A statement "$H(U) = 0.43$" is completely meaningless: since

$$\log_b \xi = \frac{\log_2 \xi}{\log_2 b}, \tag{5.15}$$

0.43 could mean anything as, e.g.,

$$\text{if } b = 2: \quad H(U) = 0.43 \text{ bits,} \tag{5.16}$$

$$\text{if } b = e: \quad H(U) = 0.43 \text{ nats} \simeq 0.620 \text{ bits,} \tag{5.17}$$

$$\text{if } b = 256 = 2^8: \quad H(U) = 0.43 \text{ ``bytes''} = 3.44 \text{ bits.} \tag{5.18}$$

Note that the term *bits* is used in two ways: its first meaning is the *unit* of entropy when the base of the logarithm is chosen to be 2; its second meaning is *binary digits*, i.e. in particular the number of digits of a binary codeword.

Remark 5.6 It is worth mentioning that if all r events are equally likely, Shannon's definition of entropy reduces to Hartley's measure:

$$p_i = \frac{1}{r}, \forall i: \quad H(U) = -\sum_{i=1}^{r} \frac{1}{r} \log_b \frac{1}{r} = \frac{1}{r} \log_b r \cdot \underbrace{\sum_{i=1}^{r} 1}_{=r} = \log_b r. \tag{5.19}$$

Remark 5.7 Be careful not to confuse *uncertainty* with *information*. For motivation purposes, in Section 5.1 we talked a lot about "information." However, what we actually meant there is "self-information" or, more nicely put, "uncertainty." You will learn in Chapter 6 that information is what you get by reducing uncertainty and see a formal definition of information there.

Another important observation is that the entropy of U does not depend on the different possible values that U can take on, but only on the *probabilities* of these values. Hence,

$$U \in \{ \underbrace{1}_{\substack{\text{with}\\ \text{prob. } \frac{1}{2}}}, \underbrace{2}_{\substack{\text{with}\\ \text{prob. } \frac{1}{3}}}, \underbrace{3}_{\substack{\text{with}\\ \text{prob. } \frac{1}{6}}} \} \tag{5.20}$$

and

$$V \in \{ \underbrace{34}_{\substack{\text{with}\\ \text{prob. } \frac{1}{2}}}, \underbrace{512}_{\substack{\text{with}\\ \text{prob. } \frac{1}{3}}}, \underbrace{981}_{\substack{\text{with}\\ \text{prob. } \frac{1}{6}}} \} \tag{5.21}$$

have both the same entropy, which is

$$H(U) = H(V) = -\frac{1}{2}\log_2 \frac{1}{2} - \frac{1}{3}\log_2 \frac{1}{3} - \frac{1}{6}\log_2 \frac{1}{6} \simeq 1.46 \text{ bits}. \tag{5.22}$$

5.2.2 Binary entropy function

One special case of entropy is so important that we introduce a specific name.

Definition 5.8 (Binary entropy function) If U is *binary* with two possible values u_1 and u_2 such that $\Pr[U = u_1] = p$ and $\Pr[U = u_2] = 1 - p$, then

$$H(U) = H_b(p), \tag{5.23}$$

where $H_b(\cdot)$ is called the *binary entropy function* and is defined as

$$H_b(p) \triangleq -p\log_2 p - (1-p)\log_2(1-p), \qquad p \in [0,1]. \tag{5.24}$$

The function $H_b(\cdot)$ is shown in Figure 5.2.

Exercise 5.9 *Show that the maximal value of $H_b(p)$ is 1 bit and is taken on for $p = 1/2$.* ◊

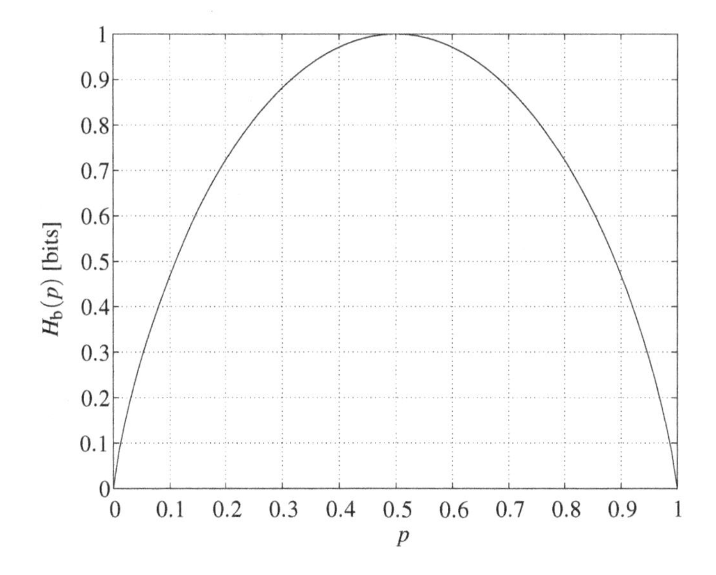

Figure 5.2 Binary entropy function $H_b(p)$ as a function of the probability p.

5.2.3 The Information Theory Inequality

The following inequality does not really have a name, but since it is so important in information theory, we will follow James Massey, retired professor at ETH in Zurich, and call it the *Information Theory Inequality* or the *IT Inequality*.

Lemma 5.10 (IT Inequality) *For any base $b > 0$ and any $\xi > 0$,*

$$\left(1 - \frac{1}{\xi}\right)\log_b e \le \log_b \xi \le (\xi - 1)\log_b e \tag{5.25}$$

with equalities on both sides if, and only if, $\xi = 1$.

Proof Actually, Figure 5.3 can be regarded as a proof. For those readers who would like a formal proof, we provide a mathematical derivation. We start with the upper bound. First note that

$$\log_b \xi \big|_{\xi=1} = 0 = (\xi - 1)\log_b e\big|_{\xi=1}. \tag{5.26}$$

Then have a look at the derivatives:

$$\frac{\mathrm{d}}{\mathrm{d}\xi}\log_b \xi = \frac{1}{\xi}\log_b e \begin{cases} > \log_b e & \text{if } 0 < \xi < 1, \\ < \log_b e & \text{if } \xi > 1, \end{cases} \tag{5.27}$$

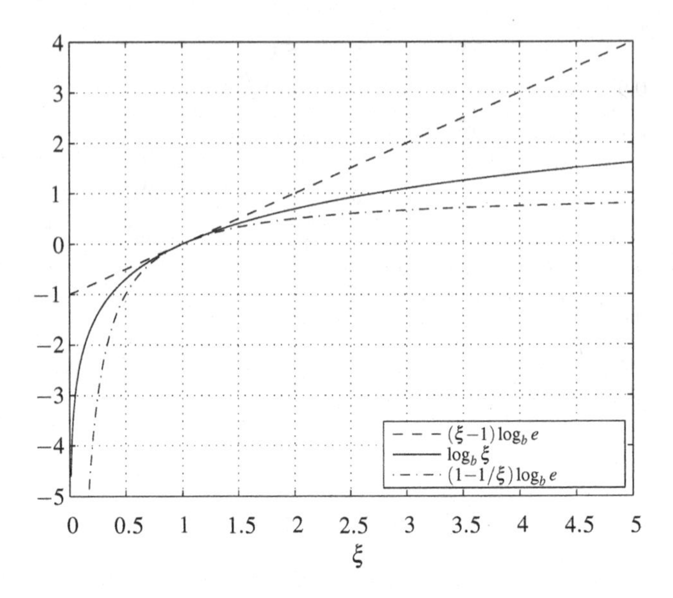

Figure 5.3 Illustration of the IT Inequality.

and

$$\frac{\mathrm{d}}{\mathrm{d}\xi}(\xi-1)\log_b e = \log_b e. \tag{5.28}$$

Hence, the two functions coincide at $\xi = 1$, and the linear function is above the logarithm for all other values.

To prove the lower bound again note that

$$\left(1-\frac{1}{\xi}\right)\log_b e\bigg|_{\xi=1} = 0 = \log_b \xi\,\big|_{\xi=1} \tag{5.29}$$

and

$$\frac{\mathrm{d}}{\mathrm{d}\xi}\left(1-\frac{1}{\xi}\right)\log_b e = \frac{1}{\xi^2}\log_b e \begin{cases} > \frac{\mathrm{d}}{\mathrm{d}\xi}\log_b \xi = \frac{1}{\xi}\log_b e & \text{if } 0<\xi<1, \\ < \frac{\mathrm{d}}{\mathrm{d}\xi}\log_b \xi = \frac{1}{\xi}\log_b e & \text{if } \xi>1, \end{cases} \tag{5.30}$$

similarly to above. □

5.2.4 Bounds on the entropy

Lemma 5.11 *If U has r possible values, then*

$$0 \le H(U) \le \log_2 r \text{ bits}, \tag{5.31}$$

where

$$H(U) = 0 \qquad \textit{if, and only if, } p_i = 1 \textit{ for some } i, \tag{5.32}$$

$$H(U) = \log_2 r \textit{ bits} \qquad \textit{if, and only if, } p_i = \frac{1}{r}\ \forall i. \tag{5.33}$$

Proof Since $0 \leq p_i \leq 1$, we have

$$-p_i \log_2 p_i \begin{cases} = 0 & \text{if } p_i = 1, \\ > 0 & \text{if } 0 < p_i < 1. \end{cases} \tag{5.34}$$

Hence, $H(U) \geq 0$. Equality can only be achieved if $-p_i \log_2 p_i = 0$ for all i, i.e. $p_i = 1$ for one i and $p_i = 0$ for the rest.

To derive the upper bound, we use a trick that is quite common in information theory: we take the difference and try to show that it must be nonpositive. In the following we arrange the probabilities in descending order and assume that r' ($r' \leq r$) of the r values of the probabilities p_i are strictly positive, i.e. $p_i > 0$ for all $i = 1, \ldots, r'$, and $p_i = 0$ for $i = r' + 1, \ldots, r$. Then

$$H(U) - \log_2 r = -\sum_{i=1}^{r} p_i \log_2 p_i - \log_2 r \tag{5.35}$$

$$= -\sum_{i=1}^{r'} p_i \log_2 p_i - \log_2 r \cdot \underbrace{\sum_{i=1}^{r'} p_i}_{=1} \tag{5.36}$$

$$= -\sum_{i=1}^{r'} p_i \log_2 p_i - \sum_{i=1}^{r'} p_i \log_2 r \tag{5.37}$$

$$= -\sum_{i=1}^{r'} p_i \log_2 (p_i \cdot r) \tag{5.38}$$

$$= \sum_{i=1}^{r'} p_i \log_2 \underbrace{\left(\frac{1}{p_i \cdot r}\right)}_{\triangleq \xi} \tag{5.39}$$

$$\leq \sum_{i=1}^{r'} p_i \left(\frac{1}{p_i \cdot r} - 1\right) \cdot \log_2 e \tag{5.40}$$

$$= \left(\sum_{i=1}^{r'} \frac{1}{r} - \underbrace{\sum_{i=1}^{r'} p_i}_{=1}\right) \cdot \log_2 e \tag{5.41}$$

$$= \left(\frac{r'}{r} - 1\right) \cdot \log_2 e \tag{5.42}$$

$$\leq (1 - 1) \cdot \log_2 e = 0. \tag{5.43}$$

Here, (5.40) follows from the IT Inequality (Lemma 5.10), and (5.43) follows because $r' \leq r$. Hence, $H(U) \leq \log_2 r$.

Equality can only be achieved if both

(1) in the IT Inequality $\xi = 1$, i.e. if $1/p_i r = 1$ for all i, i.e. if $p_i = 1/r$ for all i; and
(2) $r' = r$.

Note that if the first condition is satisfied, then the second condition is automatically satisfied. □

5.3 Trees revisited

The most elegant way to connect our new definition of entropy with the codes introduced in Chapter 4 is to rely again on trees with probabilities.

Consider a binary tree with probabilities. We remind the reader of our notation:

- n denotes the total number of leaves;
- p_i, $i = 1, \ldots, n$, denote the probabilities of the leaves;
- N denotes the number of nodes (including the root, but excluding the leaves); and
- P_ℓ, $\ell = 1, \ldots, \mathsf{N}$, denote the probabilities of the nodes, where by definition $P_1 = 1$ is the root probability.

Moreover, we will use $q_{\ell,j}$ to denote the probability of the jth node/leaf that is one step forward from node ℓ (the jth child of node ℓ), where $j = 0, 1$. That is, we have

$$q_{\ell,0} + q_{\ell,1} = P_\ell. \tag{5.44}$$

Now we give the following definitions.

Definition 5.12 The *leaf entropy* is defined as

$$H_{\text{leaf}} \triangleq -\sum_{i=1}^{n} p_i \log_2 p_i. \tag{5.45}$$

Definition 5.13 Denoting by $P_1, P_2, \ldots, P_{\mathsf{N}}$ the probabilities of all nodes (including the root) and by $q_{\ell,j}$ the probability of the nodes and leaves one step forward from node ℓ, we define the *branching entropy* H_ℓ of node ℓ as

$$H_\ell \triangleq -\frac{q_{\ell,0}}{P_\ell} \log_2 \frac{q_{\ell,0}}{P_\ell} - \frac{q_{\ell,1}}{P_\ell} \log_2 \frac{q_{\ell,1}}{P_\ell}. \tag{5.46}$$

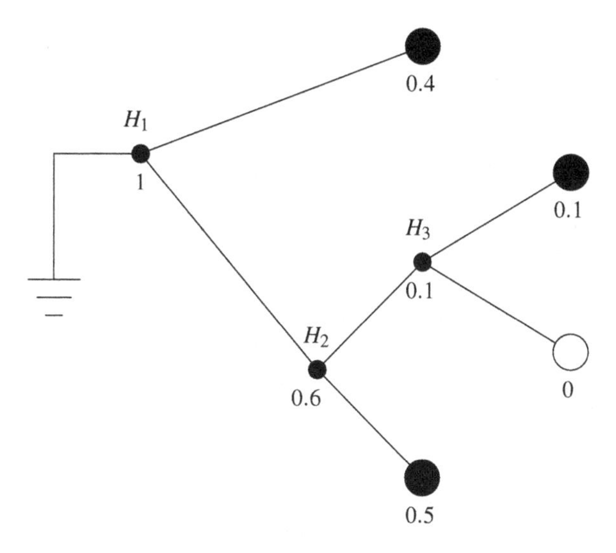

Figure 5.4 An example of a binary tree with probabilities to illustrate the calculations of the leaf entropy and the branching entropies.

Note that following Remark 5.5 we implicitly assume that the sum is only over those j for which $q_{\ell,j} > 0$, i.e. we have $H_\ell = 0$ if one of the $q_{\ell,j}$ is zero. Note further that $q_{\ell,j}/P_\ell$ is the conditional probability of going along the jth branch given that we are at node ℓ (normalization!).

Example 5.14 As an example consider the tree in Figure 5.4. We have

$$H_{\text{leaf}} = -0.4\log_2 0.4 - 0.1\log_2 0.1 - 0.5\log_2 0.5 \simeq 1.361 \text{ bits}; \tag{5.47}$$

$$H_1 = -\frac{0.4}{1}\log_2\frac{0.4}{1} - \frac{0.6}{1}\log_2\frac{0.6}{1} \simeq 0.971 \text{ bits}; \tag{5.48}$$

$$H_2 = -\frac{0.1}{0.6}\log_2\frac{0.1}{0.6} - \frac{0.5}{0.6}\log_2\frac{0.5}{0.6} \simeq 0.650 \text{ bits}; \tag{5.49}$$

$$H_3 = -\frac{0.1}{0.1}\log_2\frac{0.1}{0.1} = 0 \text{ bits}. \tag{5.50}$$

$\diamondsuit$

We will next prove a very interesting relationship between the leaf entropy and the branching entropy that will turn out to be fundamental for the understanding of codes.

Theorem 5.15 (Leaf Entropy Theorem) *In any tree with probabilities we have that*

$$H_{\text{leaf}} = \sum_{\ell=1}^{\mathrm{N}} P_\ell H_\ell. \tag{5.51}$$

Proof Recall that by the definition of trees and trees with probabilities we have, for every node ℓ,

$$P_\ell = q_{\ell,0} + q_{\ell,1}. \tag{5.52}$$

Using the definition of branching entropy, we obtain

$$\begin{aligned}
P_\ell H_\ell &= P_\ell \cdot \left(-\frac{q_{\ell,0}}{P_\ell} \log_2 \frac{q_{\ell,0}}{P_\ell} - \frac{q_{\ell,1}}{P_\ell} \log_2 \frac{q_{\ell,1}}{P_\ell} \right) &(5.53)\\
&= -q_{\ell,0} \log_2 \frac{q_{\ell,0}}{P_\ell} - q_{\ell,1} \log_2 \frac{q_{\ell,1}}{P_\ell} &(5.54)\\
&= -q_{\ell,0} \log_2 q_{\ell,0} - q_{\ell,1} \log_2 q_{\ell,1} + q_{\ell,0} \log_2 P_\ell + q_{\ell,1} \log_2 P_\ell &(5.55)\\
&= -q_{\ell,0} \log_2 q_{\ell,0} - q_{\ell,1} \log_2 q_{\ell,1} + \underbrace{(q_{\ell,0} + q_{\ell,1})}_{=P_\ell} \log_2 P_\ell &(5.56)\\
&= -q_{\ell,0} \log_2 q_{\ell,0} - q_{\ell,1} \log_2 q_{\ell,1} + P_\ell \log_2 P_\ell, &(5.57)
\end{aligned}$$

where the last equality follows from (5.52).

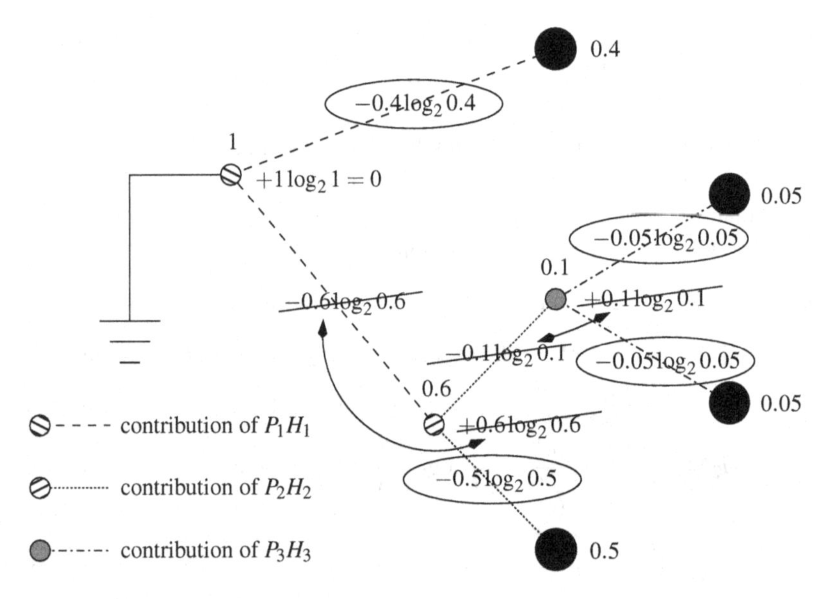

Figure 5.5 Graphical proof of the Leaf Entropy Theorem. There are three nodes: we see that all contributions cancel apart from the root node (whose contribution is 0) and the leaves.

Hence, for every node $\tilde{\ell}$, we see that it will contribute to $\sum_{\ell=1}^{N} P_\ell H_\ell$ twice:

- firstly it will add $P_{\tilde{\ell}} \log_2 P_{\tilde{\ell}}$ when the node counter ℓ passes through $\tilde{\ell}$; but
- secondly it will subtract $q_{\ell,j} \log_2 q_{\ell,j} = P_{\tilde{\ell}} \log_2 P_{\tilde{\ell}}$ when the node counter ℓ points to the parent node of $\tilde{\ell}$.

Hence, the contributions of all nodes will be canceled out – apart from the root that does not have a parent! The root only contributes $P_1 \log_2 P_1$ for $\ell = 1$. However, since $P_1 = 1$, we have $P_1 \log_2 P_1 = 1 \log_2 1 = 0$. So the root does not contribute either.

It only remains to consider the leaves. Note that the node counter ℓ will not pass through leaves by definition. Hence, a leaf only contributes when the node counter points to its parent node and its contribution is $-q_{\ell,j} \log_2 q_{\ell,j} = -p_i \log_2 p_i$. Since the sum of all $-p_i \log_2 p_i$ equals the leaf entropy by definition, this proves the claim.

In Figure 5.5 we have tried to depict this proof graphically. □

Example 5.16 (Continuation from Example 5.14) Using the values from (5.47)–(5.50) we obtain

$$P_1 H_1 + P_2 H_2 + P_3 H_3 = 1 \cdot 0.971 + 0.6 \cdot 0.650 + 0.1 \cdot 0 \text{ bits} \tag{5.58}$$

$$= 1.361 \text{ bits} = H_{\text{leaf}} \tag{5.59}$$

as expected. ◊

5.4 Bounds on the efficiency of codes

The main strength of information theory is that it can provide some fundamental statements about what is possible and what is not possible to achieve. So a typical information theoretic result will consist of an upper bound and a lower bound or, in general, an achievability part and a converse part. The achievability part of a theorem tells us what we can do, and the converse part tells us what we cannot do.

Sometimes, the theorem will also tell us how to do it, but usually the result is theoretic in the sense that it only proves what is possible without actually saying how it could be done. To put it pointedly: information theory tells us what is possible; coding theory tells us how to do it.

5.4.1 What we cannot do: fundamental limitations of source coding

Let us quickly summarize what we know about codes and their corresponding trees.

- The most efficient codes can always be chosen to be prefix-free (Theorem 4.19).

- Every prefix-free code can be represented by a tree where every codeword corresponds to a leaf in the tree.
- Every codeword has a certain probability corresponding to the probability of the symbol it represents.
- Unused leaves can be regarded as symbols that never occur, i.e. we assign probability zero to them.

Hence, from these observations we immediately see that the entropy $H(U)$ of a random message U with probabilities $p_1, \ldots, p_r$ and the leaf entropy of the corresponding tree are the same:

$$H_{\text{leaf}} = H(U). \tag{5.60}$$

Note that the unused leaves do not contribute to H_{leaf} since they have zero probability.

Moreover, the average codeword length L_{av} is equivalent to the average depth of the leaves. (Again we can ignore the unused leaves, since they have probability zero and therefore do not contribute to the average.)

Now note that since we consider binary trees where each node branches into two different children, we know from Lemma 5.11 that the branching entropy can be upper-bounded as follows:

$$H_\ell \leq \log_2 2 = 1 \text{ bit}. \tag{5.61}$$

Hence, using this together with the Leaf Entropy Theorem (Theorem 5.15) and the Path Length Lemma (Lemma 4.11) we obtain the following:

$$H(U) = H_{\text{leaf}} = \sum_{\ell=1}^{\mathsf{N}} P_\ell H_\ell \leq \sum_{\ell=1}^{\mathsf{N}} P_\ell \cdot 1 \text{ bit} = \sum_{\ell=1}^{\mathsf{N}} P_\ell = \mathsf{L}_{\text{av}} \text{ bits}. \tag{5.62}$$

In other words,

$$\mathsf{L}_{\text{av}} \geq H(U) \text{ bits}. \tag{5.63}$$

This is the *converse part* of the *Coding Theorem for a Single Random Message*. It says that whatever code you try to design, the average codeword length of *any* binary code for an r-ary random message U cannot be smaller than the entropy of U (using the correct unit of bits)!

Note that to prove this statement we have not designed any code, but instead we have been able to prove something that holds for *every* code that exists!

When do we have equality? From the above derivation we see that we have equality if the branching entropy is always 1 bit, $H_\ell = 1$ bit, i.e. the branching probabilities are all uniform. This is only possible if p_i is a negative integer power of 2 for all i:

$$p_i = 2^{-\nu_i} \tag{5.64}$$

with v_i a natural number (and, of course, if we design an optimal code).

5.4.2 What we can do: analysis of the best codes

In practice, it is not only important to know where the limitations are, but also perhaps even more so to know how close we can get to these limitations. So as a next step we would like to analyze the best codes (i.e. the Huffman codes derived in Section 4.6) and see how close they get to the limitations shown in Section 5.4.1.

Unfortunately, it is rather difficult to analyze Huffman codes. To circumvent this problem, we will design a new code, called the *Fano code*, and analyze its performance instead. Fano codes are not optimal in general, i.e. their performance is worse than the performance of Huffman codes. Therefore any upper bound on L_{av} that can be achieved by a Fano code can definitely also be achieved by a Huffman code.

Definition 5.17 (Fano code) The *Fano code*[2] is generated according to the following algorithm:

Step 1 Arrange the symbols in order of nonincreasing probability.

Step 2 Divide the list of ordered symbols into two parts, with the total probability of the left part being as close to the total probability of the right part as possible.

Step 3 Assign the binary digit 0 to the left part of the list, and the digit 1 to the right part. This means that the codewords for the symbols in the first part will all start with 0, and the codewords for the symbols in the second part will all start with 1.

Step 4 Recursively apply Step 2 and Step 3 to each of the two parts, subdividing into further parts and adding bits to the codewords until each symbol is the single member of a part.

Note that effectively this algorithm constructs a tree. Hence, the Fano code is prefix-free.

Example 5.18 Let us generate the Fano code for a random message with five symbols having probabilities

$$\begin{aligned} &p_1 = 0.35, \qquad p_2 = 0.25, \qquad p_3 = 0.15, \\ &p_4 = 0.15, \qquad p_5 = 0.1. \end{aligned} \tag{5.65}$$

[2] Note that this code is usually known as the *Shannon–Fano code*. However, this is a misnaming because it was Fano's invention. Shannon proposed a slightly different code, which unfortunately is also known as the *Shannon–Fano code*. For more details on this confusion, we refer to the discussion in Section 5.6.

Since the symbols are already ordered in decreasing order of probability, Step 1 can be omitted. We hence want to split the list into two parts, both having as similar total probability as possible. If we split $\{1\}$ and $\{2,3,4,5\}$, we have a total probability 0.35 on the left and 0.65 on the right; the split $\{1,2\}$ and $\{3,4,5\}$ yields 0.6 and 0.4; and $\{1,2,3\}$ and $\{4,5\}$ gives 0.75 and 0.25. We see that the second split is best. So we assign 0 as a first digit to $\{1,2\}$ and 1 to $\{3,4,5\}$.

Now we repeat the procedure with both subgroups. Firstly, we split $\{1,2\}$ into $\{1\}$ and $\{2\}$. This is trivial. Secondly, we split $\{3,4,5\}$ into $\{3\}$ and $\{4,5\}$ because 0.15 and 0.25 is closer than 0.3 and 0.1 that we would have obtained by dividing into $\{3,4\}$ and $\{5\}$. Again we assign the corresponding second digits.

Finally, we split the last group $\{4,5\}$ into $\{4\}$ and $\{5\}$. We end up with the five codewords $\{00,01,10,110,111\}$. This whole procedure is shown in Figure 5.6. ◊

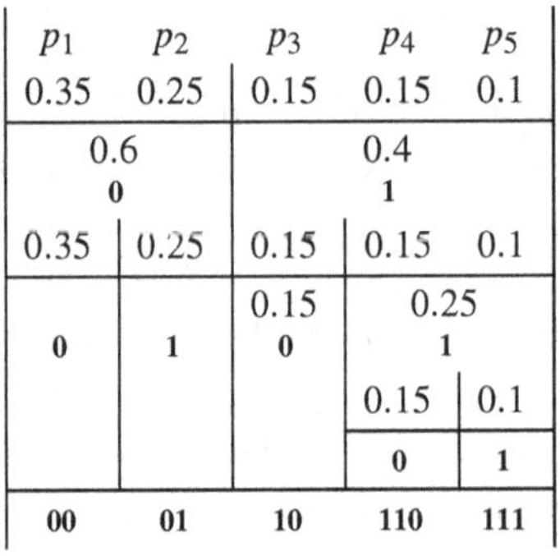

p_1	p_2	p_3	p_4	p_5
0.35	0.25	0.15	0.15	0.1
0.6		0.4		
0		1		
0.35	0.25	0.15	0.15	0.1
		0.15	0.25	
0	1	0	1	
			0.15	0.1
			0	1
00	01	10	110	111

Figure 5.6 Construction of the Fano code of Example 5.18.

Exercise 5.19 *Construct the Fano code for the random message U of Example 4.16 with four symbols having probabilities*

$$p_1 = 0.4, \qquad p_2 = 0.3, \qquad p_3 = 0.2, \qquad p_4 = 0.1, \tag{5.66}$$

and show that it is identical to the corresponding Huffman code. ◊

Remark 5.20 We would like to point out that there are cases where the algorithm given in Definition 5.17 does not lead to a unique design: there might be two different ways of dividing the list into two parts such that the total probabilities are as similar as possible. Since the algorithm does not specify what to do in such a case, you are free to choose any possible way. Unfortunately, however, these different choices can lead to codes with different performance.[3] As an example, consider a random message U with seven possible symbols having the following probabilities:

$$\begin{aligned} &p_1 = 0.35, \quad &&p_2 = 0.3, \quad &&p_3 = 0.15, \quad &&p_4 = 0.05, \\ &p_5 = 0.05, \quad &&p_6 = 0.05, \quad &&p_7 = 0.05. \end{aligned} \tag{5.67}$$

Figures 5.7 and 5.8 show two different possible Fano codes for this random message. The first has an average codeword length of $\mathsf{L}_{\text{av}} = 2.45$, while the latter's performance is better with $\mathsf{L}_{\text{av}} = 2.4$.

p_1	p_2	p_3	p_4	p_5	p_6	p_7
0.35	0.3	0.15	0.05	0.05	0.05	0.05
0.65 0		0.35 1				
0.35	0.3	0.15	0.05	0.05	0.05	0.05
0	1	0.2 0		0.15 1		
		0.15	0.05	0.05	0.05	0.05
		0	1	0.1 0		0.05 1
				0.05	0.05	
				0	1	
00	01	100	101	1100	1101	111

Figure 5.7 One possible Fano code for the random message given in (5.67).

Exercise 5.21 *In total there are six different possible designs of a Fano code for the random message given in Remark 5.20. Design all of them and compare their performances.* ◊

We next prove a simple property of the Fano code.

Lemma 5.22 *The codeword lengths l_i of a Fano code satisfy the following:*

$$l_i \le \left\lceil \log_2 \frac{1}{p_i} \right\rceil, \tag{5.68}$$

where $\lceil \xi \rceil$ denotes the smallest integer not smaller than ξ.

[3] This cannot happen in the case of a Huffman code! Even though the algorithm of the Huffman code is not unique either, it always will result in codes of equal (optimal) performance. The reason for this is clear: we have proven that the Huffman algorithm results in an *optimal* code.

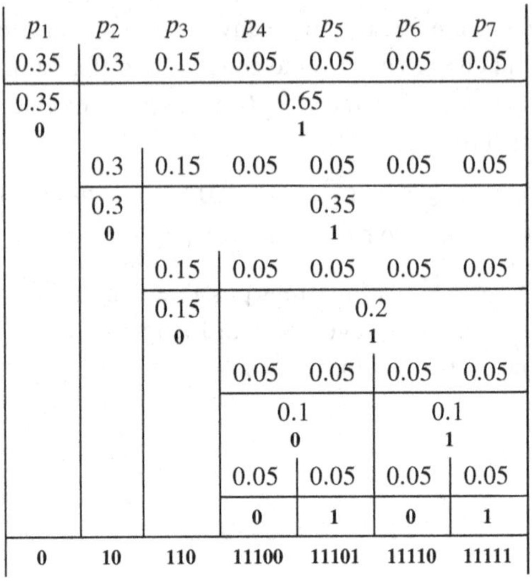

p_1	p_2	p_3	p_4	p_5	p_6	p_7
0.35	0.3	0.15	0.05	0.05	0.05	0.05
0.35	0.65					
0	1					
	0.3	0.15	0.05	0.05	0.05	0.05
	0.3	0.35				
	0	1				
		0.15	0.05	0.05	0.05	0.05
		0.15	0.2			
		0	1			
			0.05	0.05	0.05	0.05
			0.1		0.1	
			0		1	
			0.05	0.05	0.05	0.05
			0	1	0	1
0	10	110	11100	11101	11110	11111

Figure 5.8 A second possible Fano code for the random message given in (5.67).

Proof By construction, any symbol with probability $p_i \geq 1/2$ will be alone in one part in the first round of the algorithm. Hence,

$$l_i = 1 = \left\lceil \log_2 \frac{1}{p_i} \right\rceil. \tag{5.69}$$

If $1/4 \leq p_i < 1/2$, then at the latest in the second round of the algorithm the symbol will occupy one partition. (Note that it is possible that the symbol is already the single element of one partition in the first round. For example, for $p_1 = 3/4$ and $p_2 = 1/4$, p_2 will have $l_2 = 1$.) Hence, we have

$$l_i \leq 2 = \left\lceil \log_2 \frac{1}{p_i} \right\rceil. \tag{5.70}$$

In the same fashion we show that for $1/8 \leq p_i < 1/4$,

$$l_i \leq 3 = \left\lceil \log_2 \frac{1}{p_i} \right\rceil; \tag{5.71}$$

for $1/16 \leq p_i < 1/8$,

$$l_i \leq 4 = \left\lceil \log_2 \frac{1}{p_i} \right\rceil; \tag{5.72}$$

etc. □

Next, let us see how efficient the Fano code is. To that goal, we note that from (5.68) we have

$$l_i \le \left\lceil \log_2 \frac{1}{p_i} \right\rceil < \log_2 \frac{1}{p_i} + 1. \tag{5.73}$$

We get

$$\begin{aligned}
\mathsf{L}_{\text{av}} &= \sum_{i=1}^{r} p_i l_i && (5.74)\\
&< \sum_{i=1}^{r} p_i \left(\log_2 \frac{1}{p_i} + 1 \right) && (5.75)\\
&= \sum_{i=1}^{r} p_i \log_2 \frac{1}{p_i} + \sum_{i=1}^{r} p_i && (5.76)\\
&= -\sum_{i=1}^{r} p_i \log_2 p_i + 1 && (5.77)\\
&= H(U) + 1 \text{ bits}, && (5.78)
\end{aligned}$$

where the entropy is based on the binary logarithm, i.e. it is measured in bits. Hence, the Fano code (even though it is *not* an optimal code) approaches the ultimate lower bound (5.63) by less than 1 bit! A Huffman code will be even better than that.

5.4.3 Coding Theorem for a Single Random Message

We summarize this so far most important result of this chapter.

Theorem 5.23 (Coding Theorem for a Single Random Message) *For an optimal binary prefix-free code (i.e. a binary Huffman code) for an r-ary random message U, the average codeword length L_{av} satisfies*

$$H(U) \text{ bits} \le \mathsf{L}_{\text{av}} < H(U) + 1 \text{ bits} \tag{5.79}$$

(where the entropy is measured in bits). We have equality on the left if, and only if, p_i is a negative integer power of 2, $\forall i$.
Moreover, this statement also holds true for Fano coding.

Example 5.24 We consider a random message U with seven symbols having probabilities

$$\begin{aligned}
&p_1 = 0.4, \quad p_2 = 0.1, \quad p_3 = 0.1, \quad p_4 = 0.1,\\
&p_5 = 0.1, \quad p_6 = 0.1, \quad p_7 = 0.1,
\end{aligned} \tag{5.80}$$

i.e. $H(U) \simeq 2.52$ bits. We firstly design a Fano code; see Figure 5.9. The corresponding tree is shown in Figure 5.10. Note that the construction algorithm is not unique in this case: in the second round we could split the second group either to $\{3,4\}$ and $\{5,6,7\}$ or $\{3,4,5\}$ and $\{6,7\}$. In this case, both ways will result in a code of identical performance. The same situation occurs in the third round.

p_1	p_2	p_3	p_4	p_5	p_6	p_7
0.4	0.1	0.1	0.1	0.1	0.1	0.1
0.5				0.5		
0				1		
0.4	0.1	0.1	0.1	0.1	0.1	0.1
		0.2			0.3	
0	1	0			1	
		0.1	0.1	0.1	0.1	0.1
				0.2		0.1
		0	1	0		1
				0.1	0.1	
				0	1	
00	01	100	101	1100	1101	111

Figure 5.9 Construction of the Fano code of Example 5.24.

The efficiency of this Fano code is given by

$$\mathrm{L_{av}} = 1 + 0.5 + 0.5 + 0.2 + 0.3 + 0.2 = 2.7 \text{ bits}, \tag{5.81}$$

which satisfies, as predicted,

$$2.52 \text{ bits} \le 2.7 \text{ bits} < 3.52 \text{ bits}. \tag{5.82}$$

A corresponding Huffman code for U is shown in Figure 5.11. Its performance is $\mathrm{L_{av}} = 2.6$ bits, i.e. it is better than the Fano code, but of course it still holds that

$$2.52 \text{ bits} \le 2.6 \text{ bits} < 3.52 \text{ bits}. \tag{5.83}$$

◊

Exercise 5.25 *Design a Huffman code and a Fano code for the random message U with probabilities*

$$\begin{aligned} &p_1 = 0.25, && p_2 = 0.2, && p_3 = 0.2, && p_4 = 0.1, \\ &p_5 = 0.1, && p_6 = 0.1, && p_7 = 0.05, \end{aligned} \tag{5.84}$$

and compare their performances. ◊

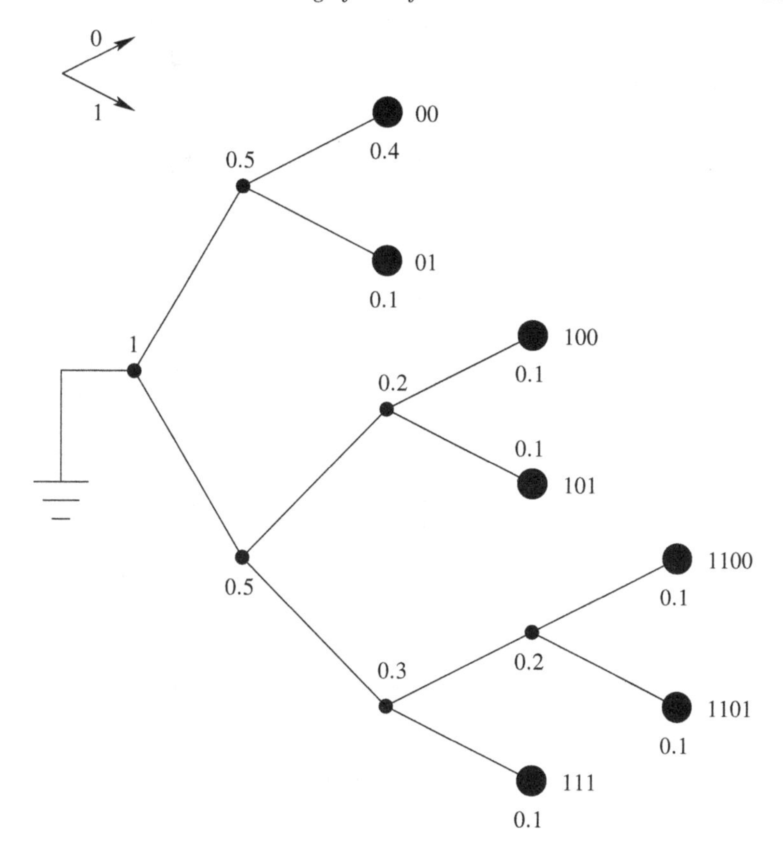

Figure 5.10 A Fano code for the message U of Example 5.24.

We have seen in the above examples and exercises that, even though the Fano code is not optimal, its performance is usually very similar to the optimal Huffman code. In particular, we know from Theorem 5.23 that the performance gap is less than one binary digit.

We are going to see next that once we start encoding not a single random message, but a sequence of such messages emitted by a random source, this difference becomes negligible.

5.5 Coding of an information source

So far we have only considered a single random message, but in reality we are much more likely to encounter a situation where we have a *stream* of messages

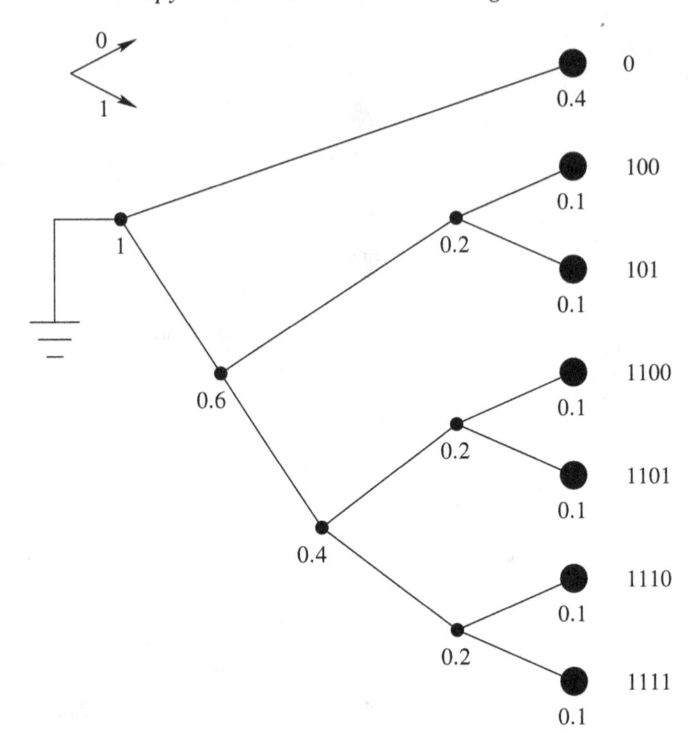

Figure 5.11 A Huffman code for the message U of Example 5.24.

that should be encoded continuously. Luckily we have prepared ourselves for this situation already by considering prefix-free codes only, which make sure that a sequence of codewords can be separated easily into the individual codewords.

In the following we will consider only the simplest case where the random source is memoryless, i.e. each symbol that is emitted by the source is independent of all past symbols. A formal definition is given as follows.

Definition 5.26 (DMS) An r-ary *discrete memoryless source (DMS)* is a device whose output is a sequence of random messages $U_1, U_2, U_3, \ldots$, where

- each U_ℓ can take on r different values with probability $p_1, \ldots, p_r$, and
- the different messages U_ℓ are independent of each other.

The obvious way of designing a compression system for such a source is to design a Huffman code for U, continuously use it for each message U_ℓ, and concatenate the codewords together. The receiver can easily separate the codewords (because the Huffman code is prefix-free) and decode them to recover the sequence of messages $\{U_\ell\}$.

However, the question is whether this is the most efficient approach. Note that it is also possible to combine two or more messages

$$(U_\ell, U_{\ell+1}, \dots, U_{\ell+\nu})$$

together and design a Huffman code for these combined messages! We will show below that this latter approach is actually more efficient. But before doing so, we need to think about such *random vector messages*.

Remark 5.27 Note that a random vector message $\mathbf{V} = (U_1, \dots, U_\nu)$ is, from the mathematical point of view, no different from any other random message: it takes on a certain finite number of different values with certain probabilities. If U_ℓ is r-ary, then $\mathbf{V}$ is r^ν-ary, but otherwise there is no fundamental difference.

We can even express the entropy of $\mathbf{V}$ as a function of the entropy of U. Let q_j denote the probability of the jth symbol of $\mathbf{V}$. Since the different messages U_ℓ are independent, we have

$$q_j = p_{i_1} \cdot p_{i_2} \cdots p_{i_\nu}, \tag{5.85}$$

where p_{i_ℓ} denotes the probability of the i_ℓth symbol of U_ℓ. Hence,

$$\begin{aligned}
H(\mathbf{V}) &= -\sum_{j=1}^{r^\nu} q_j \log_2 q_j && (5.86)\\
&= -\sum_{i_1=1}^{r} \cdots \sum_{i_\nu=1}^{r} (p_{i_1} \cdot p_{i_2} \cdots p_{i_\nu}) \log_2 (p_{i_1} \cdot p_{i_2} \cdots p_{i_\nu}) && (5.87)\\
&= -\sum_{i_1=1}^{r} \cdots \sum_{i_\nu=1}^{r} (p_{i_1} \cdot p_{i_2} \cdots p_{i_\nu}) \big(\log_2 p_{i_1} + \cdots + \log_2 p_{i_\nu}\big) && (5.88)\\
&= -\sum_{i_1=1}^{r} \cdots \sum_{i_\nu=1}^{r} p_{i_1} \cdot p_{i_2} \cdots p_{i_\nu} \cdot \log_2 p_{i_1} \\
&\quad - \cdots - \sum_{i_1=1}^{r} \cdots \sum_{i_\nu=1}^{r} p_{i_1} \cdot p_{i_2} \cdots p_{i_\nu} \cdot \log_2 p_{i_\nu} && (5.89)\\
&= -\left(\sum_{i_1=1}^{r} p_{i_1} \log_2 p_{i_1}\right) \cdot \underbrace{\left(\sum_{i_2=1}^{r} p_{i_2}\right)}_{=1} \cdots \underbrace{\left(\sum_{i_\nu=1}^{r} p_{i_\nu}\right)}_{=1} \\
&\quad - \cdots - \underbrace{\left(\sum_{i_1=1}^{r} p_{i_1}\right)}_{=1} \cdots \underbrace{\left(\sum_{i_{\nu-1}=1}^{r} p_{i_{\nu-1}}\right)}_{=1} \cdot \left(\sum_{i_\nu=1}^{r} p_{i_\nu} \log_2 p_{i_\nu}\right) && (5.90)
\end{aligned}$$

$$= -\left(\sum_{i_1=1}^{r} p_{i_1} \log_2 p_{i_1}\right) - \cdots - \left(\sum_{i_\nu=1}^{r} p_{i_\nu} \log_2 p_{i_\nu}\right) \tag{5.91}$$

$$= H(U_1) + \cdots + H(U_\nu) \tag{5.92}$$

$$= \nu H(U). \tag{5.93}$$

Here the last equality follows because the entropy of all U_ℓ is identical.

In other words, since $\mathbf{V}$ consists of ν *independent* random messages U, its uncertainty is simply ν times the uncertainty of U.

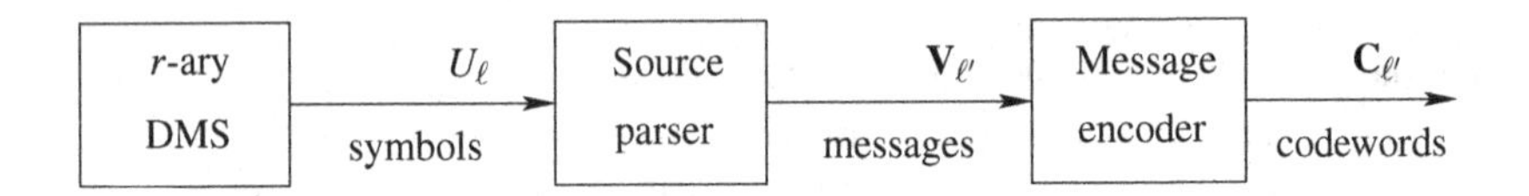

Figure 5.12 A coding scheme for an information source: the source parser groups the source output sequence $\{U_\ell\}$ into messages $\{\mathbf{V}_{\ell'}\}$. The message encoder then assigns a codeword $\mathbf{C}_{\ell'}$ to each possible message $\mathbf{V}_{\ell'}$.

Now our compression system looks as shown in Figure 5.12. The *source parser* is a device that groups ν incoming source symbols $(U_1, \ldots, U_\nu)$ together to a new message $\mathbf{V}$. Note that because the source $\{U_\ell\}$ is memoryless, the different messages $\{\mathbf{V}_{\ell'}\}$ are independent. Therefore we only need to look at one such message $\mathbf{V}$ (where we omit the time index ℓ').

So let us now use an optimal code (i.e. a Huffman code) or at least a good code (e.g. a Fano code) for the message $\mathbf{V}$. Then from the Coding Theorem for a Single Random Message (Theorem 5.23) we know that

$$H(\mathbf{V}) \text{ bits} \leq \mathsf{L}_{\text{av}} < H(\mathbf{V}) + 1 \text{ bits}, \tag{5.94}$$

where L_{av} denotes the average codeword length for the codewords describing the *vector* messages $\mathbf{V}$.

Next note that it is not really fair to compare different L_{av} because, for larger ν, L_{av} also will be larger. So, to be correct we should compute the average codeword length necessary to describe *one* source symbol. Since $\mathbf{V}$ contains ν source symbols U_ℓ, the correct measure of performance is $\mathsf{L}_{\text{av}}/\nu$.

Hence, we divide the whole expression (5.94) by ν:

$$\frac{H(\mathbf{V})}{\nu} \text{ bits} \leq \frac{\mathsf{L}_{\text{av}}}{\nu} < \frac{H(\mathbf{V})}{\nu} + \frac{1}{\nu} \text{ bits}, \tag{5.95}$$

and make use of (5.93),

$$\frac{\nu H(U)}{\nu} \text{ bits} \leq \frac{\mathsf{L}_{\text{av}}}{\nu} < \frac{\nu H(U)}{\nu} + \frac{1}{\nu} \text{ bits}; \tag{5.96}$$

i.e.,

$$H(U) \text{ bits} \leq \frac{\mathsf{L}_{\text{av}}}{\nu} < H(U) + \frac{1}{\nu} \text{ bits}. \tag{5.97}$$

Note that again we assume that the entropies are measured in bits.

We immediately get the following main result, also known as *Shannon's Source Coding Theorem.*

Theorem 5.28 (Coding Theorem for a DMS)
There exists a binary prefix-free code of a ν-block message from a discrete memoryless source (DMS) such that the average number $\mathsf{L}_{\text{av}}/\nu$ of binary code digits per source letter satisfies

$$\frac{\mathsf{L}_{\text{av}}}{\nu} < H(U) + \frac{1}{\nu} \text{ bits}, \tag{5.98}$$

where $H(U)$ is the entropy of a single source letter measured in bits. Conversely, for every *binary code of a ν-block message,*

$$\frac{\mathsf{L}_{\text{av}}}{\nu} \geq H(U) \text{ bits}. \tag{5.99}$$

Note that everywhere we need to choose the units of the entropy to be in bits.

We would like to discuss this result briefly. The main point to note here is that by choosing ν large enough, we can approach the ultimate limit of compression $H(U)$ *arbitrarily closely* when using a Huffman or a Fano code. Hence, the entropy $H(U)$ is the amount of information that is packed in the output of the discrete memoryless source U! In other words, we can compress any DMS to $H(U)$ bits on average, but not less. This is the first real justification of the usefulness of Definition 5.4.

We also see that in the end it does not make much difference whether we use a Huffman code or a suboptimal Fano code as both approach the ultimate limit for ν large enough.

On the other hand, note the price we have to pay: by making ν large, we not only increase the number of possible messages, and thereby make the code complicated, but also we introduce *delay* into the system as the encoder can only encode the message *after it has received a complete block of ν source symbols*! Basically, the more closely we want to approach the ultimate limit of entropy, the larger is our potential delay in the system.

We would like to mention that our choice of a source parser that splits the source sequence into blocks of equal length is not the only choice. It is actually possible to design source parsers that will choose blocks of varying length depending on the arriving source symbols and their probabilities. By trying to combine more likely source symbols to larger blocks, while less likely symbols are grouped to smaller blocks, we can further improve on the compression rate of our system. A parser that is optimal in a specific sense is the so-called *Tunstall source parser* [Tun67], [Mas96]. The details are outside the scope of this introduction. However, note that whatever source parser and whatever message encoder we choose, we can never beat the lower bound in (5.99).

All the systems we have discussed here contain one common drawback: we always have assumed that the probability statistics of the source is known in advance when designing the system. In a practical situation this is often not the case. What is the probability distribution of a digitized speech in a telephone system? Or of English ASCII text in comparison to French ASCII text?[4] Or of different types of music? A really practical system should work independently of the source; i.e., it should estimate the probabilities of the source symbols on the fly and adapt to it automatically. Such a system is called a *universal compression scheme*. Again, the details are outside of the scope of this introduction, but we would like to mention that such schemes exist and that commonly used compression algorithms like, e.g., ZIP successfully implement such schemes.

5.6 Some historical background

The Fano code is in the literature usually known as the *Shannon–Fano code*, even though it is an invention of Professor Robert Fano from MIT [Fan49] and not of Shannon. To make things even worse, there exists another code that is also known as the *Shannon–Fano code*, but actually should be called the *Shannon code* because it was proposed by Shannon [Sha48, Sec. 9]: the construction of the Shannon code also starts with the ordering of the symbols according to decreasing probability. The ith codeword with probability p_i is then obtained by writing the cumulative probability

$$F_i \triangleq \sum_{j=1}^{i-1} p_j \tag{5.100}$$

in binary form. For example, $F_i = 0.625$ in binary form is .101 yielding a codeword 101, or $F_i = 0.3125$ is written in binary as .0101, which then results

[4] Recall the definition of ASCII given in Table 2.2.

in a codeword 0101. Since in general this binary expansion might be infinitely long, Shannon gave the additional rule that the expansion shall be carried out to exactly l_i positions, where

$$l_i \triangleq \left\lceil \log_2 \frac{1}{p_i} \right\rceil. \tag{5.101}$$

So if $F_i = 0.625$ (with binary form .101) and p_i is such that $l_i = 5$, then the resulting codeword is 10100, or if $F_i = 0.6$, which in binary form is

$$.10011001100\ldots,$$

and p_i is such that $l_i = 3$, then the resulting codeword is 100.

It is not difficult to show that this code is prefix-free. In particular it is straightforward to show that the Kraft Inequality (Theorem 4.8) is satisfied:

$$\sum_{i=1}^{r} 2^{-l_i} = \sum_{i=1}^{r} 2^{-\left\lceil \log_2 \frac{1}{p_i} \right\rceil} \le \sum_{i=1}^{r} 2^{-\log_2 \frac{1}{p_i}} = \sum_{i=1}^{r} p_i = 1, \tag{5.102}$$

where we have used that

$$l_i = \left\lceil \log_2 \frac{1}{p_i} \right\rceil \ge \log_2 \frac{1}{p_i}. \tag{5.103}$$

Shannon's code performs similarly to the Fano code of Definition 5.17, but Fano's code is in general slightly better, as can be seen by the fact that in (5.68) we have an inequality while in (5.101) we have, by definition, equality always. However – and that is probably one of the reasons[5] why the two codes are mixed up and both are known under the same name *Shannon–Fano code* – both codes satisfy the Coding Theorem for a Single Random Message (Theorem 5.23).

Actually, one also finds that *any* code that satisfies (5.101) is called a *Shannon–Fano code*! And to complete the confusion, sometimes the Shannon–Fano code is also known as *Shannon–Fano–Elias code* [CT06, Sec. 5.9]. The reason is that the Shannon code was the origin of *arithmetic coding*, which is an elegant and efficient extension of Shannon's idea, applied to the compression of the output sequence of a random source. It is based on the insight that it is not necessary to order the output sequences $\mathbf{u}$ according to their probabilities, but that it is sufficient to have them ordered lexicographically (according to the alphabet). The codeword for $\mathbf{u}$ is then the truncated binary form of the

[5] Another reason is that Shannon, when introducing his code in [Sha48, Sec. 9], also refers to Fano's code construction.

cumulative probability

$$F_{\mathbf{u}} \triangleq \sum_{\substack{\text{all sequences } \tilde{\mathbf{u}} \\ \text{that are alphabetically} \\ \text{before } \mathbf{u}}} p_{\tilde{\mathbf{u}}}, \tag{5.104}$$

where $p_{\tilde{\mathbf{u}}}$ is the probability of $\tilde{\mathbf{u}}$. However, in order to guarantee that the code is prefix-free and because the output sequences are ordered lexicographically and not according to probability, it is necessary to increase the codeword length by 1; i.e., for arithmetic coding we have the rule that the codeword length is

$$l_{\mathbf{u}} \triangleq \left\lceil \log_2 \frac{1}{p_{\mathbf{u}}} \right\rceil + 1. \tag{5.105}$$

Note further that, since the sequences are ordered lexicographically, it is also possible to compute the cumulative probability (5.104) of a particular source output sequence $\mathbf{u}$ iteratively without having to know the probabilities of all other source sequences. These ideas have been credited to the late Professor Peter Elias from MIT (hence the name *Shannon–Fano–Elias coding*), but actually Elias denied this. The concept has probably come from Shannon himself during a talk that he gave at MIT.

For an easy-to-read introduction to arithmetic coding including its history, the introductory chapter of [Say99] is highly recommended.

5.7 Further reading

For more information about data compression, the lecture notes [Mas96] of Professor James L. Massey from ETH, Zurich, are highly recommended. They read very easily and are very precise. The presentation of the material in Chapters 4 and 5 is strongly inspired by these notes. Besides the generalization of Huffman codes to D-ary alphabets and the variable-length–to–block Tunstall codes, one also finds there details of a simple, but powerful, universal data compression scheme called *Elias–Willems coding*.

For the more commonly used *Lempel–Ziv universal compression scheme* we refer to [CT06]. This is also a good place to learn more about entropy and its properties.

One of the best books on information theory is by Robert Gallager [Gal68]; however, it is written at an advanced level. It is fairly old and therefore does not cover more recent discoveries, but it gives a very deep treatment of the foundations of information theory.

5.8 Appendix: Uniqueness of the definition of entropy

In Section 5.1 we tried to motivate the definition of entropy. Even though we partially succeeded, we were not able to provide a full justification of Definition 5.4. While Shannon did provide a mathematical justification [Sha48, Sec. 6], he did not consider it very important. We omit Shannon's argument, but instead we will now quickly summarize a slightly different result that was presented in 1956 by Aleksandr Khinchin. Khinchin specified four properties that entropy is supposed to have and then proved that, given these four properties, (5.10) is the only possible definition.

We define $H_r(p_1,\ldots,p_r)$ to be a function of r probabilities $p_1,\ldots,p_r$ that sum up to 1:

$$\sum_{i=1}^{r} p_i = 1. \tag{5.106}$$

We ask this function to satisfy the following four properties.

(1) For any r, $H_r(p_1,\ldots,p_r)$ is continuous (i.e. a slight change to the values of p_i will only cause a slight change to H_r) and symmetric in $p_1,\ldots,p_r$ (i.e. changing the order of the probabilities does not affect the value of H_r).

(2) Any event of probability zero does not contribute to H_r:

$$H_{r+1}(p_1,\ldots,p_r,0) = H_r(p_1,\ldots,p_r). \tag{5.107}$$

(3) H_r is maximized by the uniform distribution:

$$H_r(p_1,\ldots,p_r) \leq H_r\left(\frac{1}{r},\ldots,\frac{1}{r}\right). \tag{5.108}$$

(4) If we partition the $m \cdot r$ possible outcomes of a random experiment into m groups, each group containing r elements, then we can do the experiment in two steps:

   (i) determine the group to which the actual outcome belongs,
   (ii) find the outcome in this group.

   Let $p_{j,i}$, $1 \leq j \leq m$, $1 \leq i \leq r$, be the probabilities of the outcomes in this random experiment. Then the total probability of all outcomes in group j is given by

$$q_j = \sum_{i=1}^{r} p_{j,i}, \tag{5.109}$$

   and the conditional probability of outcome i from group j is then given by

$$\frac{p_{j,i}}{q_j}. \tag{5.110}$$

Now $H_{m\cdot r}$ can be written as follows:

$$H_{m\cdot r}(p_{1,1},p_{1,2},\ldots,p_{m,r}) = H_m(q_1,\ldots,q_m) + \sum_{j=1}^{m} q_j H_r\left(\frac{p_{j,1}}{q_j},\ldots,\frac{p_{j,r}}{q_j}\right); \quad (5.111)$$

i.e., the uncertainty can be split into the uncertainty of choosing a group and the uncertainty of choosing one particular outcome of the chosen group, averaged over all groups.

Theorem 5.29 *The only functions H_r that satisfy the above four conditions are of the form*

$$H_r(p_1,\ldots,p_r) = -c\sum_{i=1}^{r} p_i \log_2 p_i, \quad (5.112)$$

where the constant $c > 0$ decides about the units of H_r.

Proof This theorem was proven by Aleksandr Khinchin in 1956, i.e. after Shannon had defined entropy. The article was first published in Russian [Khi56], and then in 1957 it was translated into English [Khi57]. We omit the details. □

References

[CT06] Thomas M. Cover and Joy A. Thomas, *Elements of Information Theory*, 2nd edn. John Wiley & Sons, Hoboken, NJ, 2006.

[Fan49] Robert M. Fano, "The transmission of information," Research Laboratory of Electronics, Massachusetts Institute of Technology (MIT), Technical Report No. 65, March 17, 1949.

[Gal68] Robert G. Gallager, *Information Theory and Reliable Communication.* John Wiley & Sons, New York, 1968.

[Har28] Ralph Hartley, "Transmission of information," *Bell System Technical Journal*, vol. 7, no. 3, pp. 535–563, July 1928.

[Khi56] Aleksandr Y. Khinchin, "On the fundamental theorems of information theory," (in Russian), *Uspekhi Matematicheskikh Nauk XI*, vol. 1, pp. 17–75, 1956.

[Khi57] Aleksandr Y. Khinchin, *Mathematical Foundations of Information Theory.* Dover Publications, New York, 1957.

[Mas96] James L. Massey, *Applied Digital Information Theory I and II,* Lecture notes, Signal and Information Processing Laboratory, ETH Zurich, 1995/1996. Available: http://www.isiweb.ee.ethz.ch/archive/massey_scr/

[Say99] Jossy Sayir, "On coding by probability transformation," Ph.D. dissertation, ETH Zurich, 1999, Diss. ETH No. 13099. Available: http://e-collection.ethbib.ethz.ch/view/eth:23000

[Sha37] Claude E. Shannon, "A symbolic analysis of relay and switching circuits," Master's thesis, Massachusetts Institute of Technology (MIT), August 1937.

[Sha48] Claude E. Shannon, "A mathematical theory of communication," *Bell System Technical Journal*, vol. 27, pp. 379–423 and 623–656, July and October 1948. Available: http://moser.cm.nctu.edu.tw/nctu/doc/shannon1948.pdf

[Tun67] Brian P. Tunstall, "Synthesis of noiseless compression codes," Ph.D. dissertation, Georgia Institute of Technology, September 1967.

6

Mutual information and channel capacity

6.1 Introduction

We go on to take a closer look at a typical problem in communications: how to send information reliably over a noisy communication channel. A communication channel can be thought of as the medium through which the message signal propagates from the transmit end (the source) to the receive end (the destination). A channel is said to be noisy if the data read at the channel output is not necessarily the same as the input (due to, e.g., perturbation caused by the ambient noise). Consider for example that Alice writes down a "7" on a paper with a small font size, and uses a fax machine to transfer this page to Bob. Due to limited resolution of electronic cable data transfer, Bob sees a "distorted 7" on the faxed page and could decode it incorrectly as, say, "9" (see Figure 6.1).

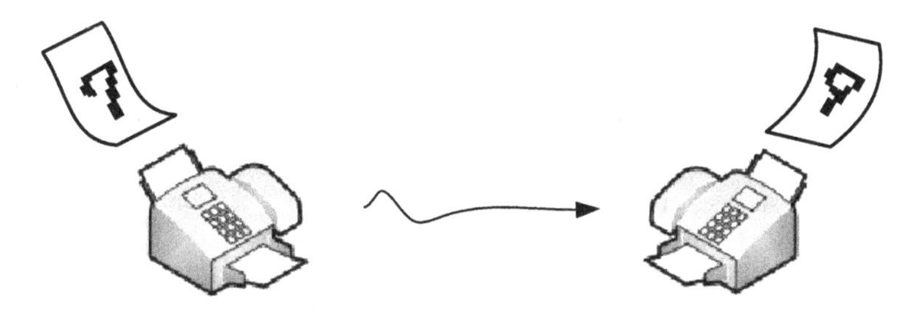

Figure 6.1 Cable data transfer as a channel.

In this example the route of data transfer through the cable acts as the channel, which is noisy since it distorts the input alphabet and, in turn, leads to possibly incorrect message decoding at the destination. It is still probable that Bob reads the message correctly as "7." The higher the probability of correct message decoding is, the more reliable the communication will be.

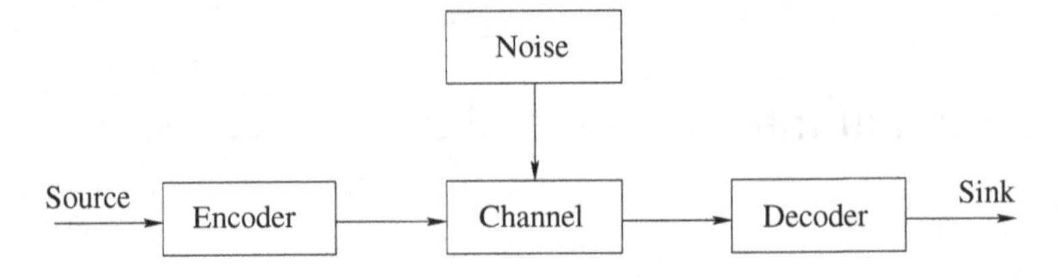

Figure 6.2 Signaling model.

The general block diagram of a typical communication system is depicted in Figure 6.2.

For a probabilistic message source, we are now able to quantify the amount of its information content in terms of the entropy defined in Chapter 5. We implicitly assume that the message has been compressed in order to remove the inherent redundancy, if any; this can be done via data compression as introduced in Chapter 4 (see also the discussion in Appendix 7.7). To combat the detrimental effect induced by the channel, the source message is further encoded with certain channel coding schemes, like the Hamming code introduced in Chapter 3. The encoded data stream is then sent over the channel. Message decoding is performed at the receiver based on the channel output. We examine each of the following problems.

- How should we measure the amount of information that can get through the channel, and what is the maximal amount?
- How can we use the channel to convey information reliably?

Note that if the channel is noiseless, i.e. the input is always reproduced at the output without errors, the answers to the aforementioned problems are simple: the maximal amount of information that can be conveyed over the channel equals the source entropy, and this can be done without any data protection mechanisms such as channel coding. If the channel is noisy, the answers turn out to be rather nontrivial. Let us begin the discussions with the mathematical model of a noisy communication channel.

6.2 The channel

Recall that in the example depicted in Figure 6.1, the input letter "7" can be either correctly decoded or mis-recognized as some other letter. The uncertainty in source symbol recovery naturally suggests a probabilistic characterization of the input–output relation of a noisy channel; such a mathematical channel model is needed in order to pin down various intrinsic properties of a channel, e.g. how much information can go through a channel.

Below is the formal definition for a channel.

Definition 6.1 (Channel) A *channel* $(\mathcal{X}, P_{Y|X}(y_j|x_i), \mathcal{Y})$ is given by

(1) an input alphabet $\mathcal{X} \triangleq \{x_1, \ldots, x_s\}$, where s denotes the number of input letters;
(2) an output alphabet $\mathcal{Y} \triangleq \{y_1, \ldots, y_t\}$, where t denotes the number of output letters; and
(3) a conditional probability distribution $P_{Y|X}(y_j|x_i)$, which specifies the probability of observing $Y = y_j$ at the output given that $X = x_i$ is sent, $1 \le i \le s$, $1 \le j \le t$.

Hence a channel with input $X \in \mathcal{X}$ and output $Y \in \mathcal{Y}$ is entirely specified by a set of conditional probabilities $P_{Y|X}(y_j|x_i)$. The size of the input and output alphabets, namely s and t, need not be the same. A schematic description of the channel is shown Figure 6.3.

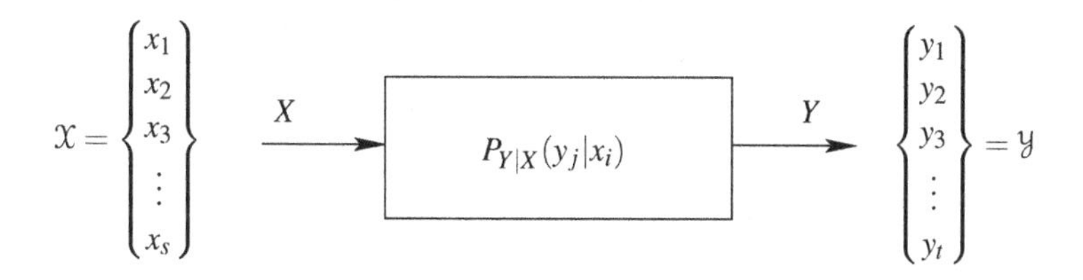

Figure 6.3 Channel model.

In this model the channel is completely described by the matrix of conditional probabilities, the so-called *channel transition matrix*:

$$\begin{pmatrix} P_{Y|X}(y_1|x_1) & P_{Y|X}(y_2|x_1) & \cdots & P_{Y|X}(y_t|x_1) \\ P_{Y|X}(y_1|x_2) & P_{Y|X}(y_2|x_2) & \cdots & P_{Y|X}(y_t|x_2) \\ \vdots & \vdots & \ddots & \vdots \\ P_{Y|X}(y_1|x_s) & P_{Y|X}(y_2|x_s) & \cdots & P_{Y|X}(y_t|x_s) \end{pmatrix}. \tag{6.1}$$

The channel transition matrix has the following properties.

(1) The entries on the ith row consist of the probabilities of observing output letters $y_1, \ldots, y_t$ given that the ith input symbol x_i is sent.
(2) The entries on the jth column consist of the probabilities of observing the jth output letter y_j given, respectively, the ith input symbols x_i are sent, $i = 1, \ldots, s$.

(3) The sum of the entries in a row is always 1, i.e.

$$\sum_{j=1}^{t} P_{Y|X}(y_j|x_i) = 1. \tag{6.2}$$

This merely means that for each input x_i we are certain that something will come out, and that the $P_{Y|X}(y_j|x_i)$ give the distribution of these probabilities.

(4) If $P_X(x_i)$ is the probability of the input symbol x_i, then

$$\sum_{i=1}^{s}\sum_{j=1}^{t} P_{Y|X}(y_j|x_i)P_X(x_i) = 1, \tag{6.3}$$

meaning that when something is put into the system, then certainly something comes out.

The probabilities $P_{Y|X}(y_j|x_i)$, $1 \leq i \leq s$, $1 \leq j \leq t$, characterize the channel completely. We assume that the channel is *stationary*, i.e. the probabilities do not change with time. We note that X is not a source but is an information-carrying channel input, which is typically a stream of encoded data (see Figure 6.2; see Chapters 3 and 7 for more details).

6.3 The channel relationships

At the transmit end we have s possible input symbols $\{x_1,\ldots,x_s\}$. If the ith symbol x_i is selected and sent over the channel, the probability of observing the jth channel output letter y_j is given by the conditional probability $P_{Y|X}(y_j|x_i)$. This means that the probability that the input–output pair (x_i,y_j) simultaneously occurs, i.e. the joint probability of $X = x_i$ and $Y = y_j$, is given by

$$P_{X,Y}(x_i,y_j) \triangleq P_{Y|X}(y_j|x_i)P_X(x_i). \tag{6.4}$$

Let us go one step further by asking the question of how to determine the probability that the jth letter y_j will occur at the channel output, hereafter denoted by $P_Y(y_j)$. A simple argument, taking into account that each input symbol occurs with probability $P_X(x_i)$, yields

$$P_Y(y_j) = P_{Y|X}(y_j|x_1)P_X(x_1) + \cdots + P_{Y|X}(y_j|x_s)P_X(x_s) \tag{6.5}$$

$$= \sum_{i=1}^{s} P_{Y|X}(y_j|x_i)P_X(x_i), \qquad 1 \leq j \leq t. \tag{6.6}$$

The above "channel equation" characterizes the input–output relation of a channel. Note that in terms of the joint probability $P_{X,Y}(x_i,y_j)$ in (6.4), we can rewrite (6.6) in a more compact form:

$$P_Y(y_j) = \sum_{i=1}^{s} P_{X,Y}(x_i, y_j), \qquad 1 \le j \le t. \tag{6.7}$$

Now take a further look at (6.4), which relates the probability of a joint occurrence of the symbol pair (x_i, y_j) with the input distribution via the *forward conditional probability* $P_{Y|X}(y_j|x_i)$ (starting from the input front with x_i given and expressing the probability that y_j is the resultant output). We can alternatively write $P_{X,Y}(x_i, y_j)$ as

$$P_{X,Y}(x_i, y_j) = P_{X|Y}(x_i|y_j) P_Y(y_j), \tag{6.8}$$

which evaluates the joint probability $P_{X,Y}(x_i, y_j)$ based on the output distribution and the *backward conditional probability* $P_{X|Y}(x_i|y_j)$ (given that y_j is received, the probability that x_i is sent). Equating (6.4) with (6.8) yields

$$P_{X|Y}(x_i|y_j) = \frac{P_{Y|X}(y_j|x_i) P_X(x_i)}{P_Y(y_j)}, \tag{6.9}$$

which is the well known *Bayes' Theorem* on conditional probabilities [BT02].

In the Bayes' formula (6.9) we can write $P_Y(y_j)$ in the denominator in terms of (6.6) to get the equivalent expression

$$P_{X|Y}(x_i|y_j) = \frac{P_{Y|X}(y_j|x_i) P_X(x_i)}{\sum_{i'=1}^{s} P_{Y|X}(y_j|x_{i'}) P_X(x_{i'})}. \tag{6.10}$$

Summing (6.10) over all the x_i clearly gives

$$\sum_{i=1}^{s} P_{X|Y}(x_i|y_j) = \sum_{i=1}^{s} \frac{P_{Y|X}(y_j|x_i) P_X(x_i)}{\sum_{i'=1}^{s} P_{Y|X}(y_j|x_{i'}) P_X(x_{i'})} \tag{6.11}$$

$$= \frac{\sum_{i=1}^{s} P_{Y|X}(y_j|x_i) P_X(x_i)}{\sum_{i'=1}^{s} P_{Y|X}(y_j|x_{i'}) P_X(x_{i'})} \tag{6.12}$$

$$= 1, \tag{6.13}$$

which means that, given output y_j, some x_i was certainly put into the channel.

6.4 The binary symmetric channel

A simple special case of a channel is the *binary channel*, which has two input symbols, 0 and 1, and two output symbols, 0 and 1; a schematic description is depicted in Figure 6.4.

The binary channel is said to be *symmetric* if

$$P_{Y|X}(0|0) = P_{Y|X}(1|1), \qquad P_{Y|X}(0|1) = P_{Y|X}(1|0). \tag{6.14}$$

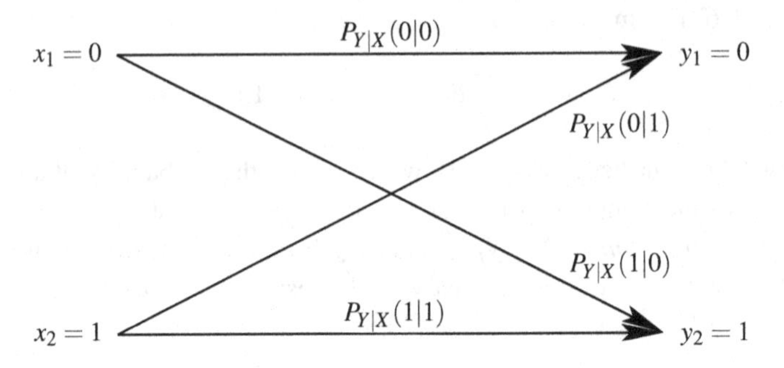

Figure 6.4 The binary channel.

Usually we abbreviate *binary symmetric channel* to *BSC*.

Let the probabilities of the input symbols be

$$P_X(0) = \delta, \tag{6.15}$$
$$P_X(1) = 1 - \delta, \tag{6.16}$$

and let the BSC probabilities be

$$P_{Y|X}(0|0) = P_{Y|X}(1|1) = 1 - \varepsilon, \tag{6.17}$$
$$P_{Y|X}(1|0) = P_{Y|X}(0|1) = \varepsilon. \tag{6.18}$$

The channel matrix is therefore

$$\begin{pmatrix} 1-\varepsilon & \varepsilon \\ \varepsilon & 1-\varepsilon \end{pmatrix} \tag{6.19}$$

and the channel relationships (6.6) become

$$P_Y(0) = (1-\varepsilon)\delta + \varepsilon(1-\delta), \tag{6.20}$$
$$P_Y(1) = \varepsilon\delta + (1-\varepsilon)(1-\delta). \tag{6.21}$$

Note that these equations can be simply checked by computing their sum:

$$P_Y(0) + P_Y(1) = (1-\varepsilon+\varepsilon)\delta + (1-\varepsilon+\varepsilon)(1-\delta) = \delta + 1 - \delta = 1. \tag{6.22}$$

Given that we know what symbol we received, what are the probabilities for the various symbols that might have been sent?

We first compute the two denominators in Equation (6.10):

$$\sum_{i=1}^{2} P_{Y|X}(y_1|x_i)P_X(x_i) = (1-\varepsilon)\delta + \varepsilon(1-\delta), \tag{6.23}$$

$$\sum_{i=1}^{2} P_{Y|X}(y_2|x_i)P_X(x_i) = \varepsilon\delta + (1-\varepsilon)(1-\delta), \tag{6.24}$$

which of course are the same as (6.20) and (6.21). We then have

$$P_{X|Y}(0|0) = \frac{(1-\varepsilon)\delta}{(1-\varepsilon)\delta + \varepsilon(1-\delta)}, \tag{6.25}$$

$$P_{X|Y}(1|0) = \frac{\varepsilon(1-\delta)}{(1-\varepsilon)\delta + \varepsilon(1-\delta)}, \tag{6.26}$$

$$P_{X|Y}(0|1) = \frac{\varepsilon\delta}{\varepsilon\delta + (1-\varepsilon)(1-\delta)}, \tag{6.27}$$

$$P_{X|Y}(1|1) = \frac{(1-\varepsilon)(1-\delta)}{\varepsilon\delta + (1-\varepsilon)(1-\delta)}. \tag{6.28}$$

Note that this involves the choice of the probabilities of the channel input.

In the special case of equally likely input symbols ($\delta = 1/2$) we have the very simple equations

$$P_{X|Y}(0|0) = P_{X|Y}(1|1) = 1-\varepsilon, \tag{6.29}$$

$$P_{X|Y}(1|0) = P_{X|Y}(0|1) = \varepsilon. \tag{6.30}$$

As a more peculiar example, suppose that $1-\varepsilon = 9/10$ and $\varepsilon = 1/10$ for the BSC, but suppose also that the probability of the input $x = 0$ being sent is $\delta = 19/20$ and $x = 1$ being sent is $1-\delta = 1/20$. We then have

$$P_{X|Y}(0|0) = \frac{171}{172}, \tag{6.31}$$

$$P_{X|Y}(1|0) = \frac{1}{172}, \tag{6.32}$$

$$P_{X|Y}(0|1) = \frac{19}{28}, \tag{6.33}$$

$$P_{X|Y}(1|1) = \frac{9}{28}. \tag{6.34}$$

Thus if we receive $y = 0$, it is more likely that $x = 0$ was sent because $171/172$ is much larger than $1/172$. If, however, $y = 1$ is received, we still have $19/28 > 9/28$, and hence $x = 0$ has a higher probability of being the one that has been sent. Therefore, $x = 0$ is always claimed regardless of the symbol received. As a result, if a stream of n binary bits is generated according to the probability law $P_X(0) = \delta = 19/20$, then there are about $n(1-\delta) = n/20$ 1s in the input sequence that will be decoded incorrectly as 0s at the channel output. Hence, the above transmission scheme does not use the channel properly, since irrespective of n it will incur an average decoding error of about $1/20$, significantly away from zero.

This situation arises whenever both following conditions are valid:

$$P_{X|Y}(0|0) > P_{X|Y}(1|0), \tag{6.35}$$

$$P_{X|Y}(0|1) > P_{X|Y}(1|1). \tag{6.36}$$

From (6.25)–(6.28) we see that for the binary symmetric channel these conditions are

$$(1-\varepsilon)\delta > \varepsilon(1-\delta), \tag{6.37}$$

$$\varepsilon\delta > (1-\varepsilon)(1-\delta), \tag{6.38}$$

or equivalently

$$\delta > \varepsilon, \tag{6.39}$$

$$\delta > 1-\varepsilon; \tag{6.40}$$

i.e., the bias in the choice of input symbols is greater than the bias of the channel. This discussion shows that it is possible, for given values of the channel transition probabilities, to come up with values for the channel input probabilities that do not make much sense in practice. As will be shown below, we can improve this if we can learn more about the fundamental characteristics of the channel and then use the channel properly through a better assignment of the input distribution.[1] To this end, we leverage the entropy defined in Chapter 5 to define the notion of "capability of a channel for conveying information" in a precise fashion.

6.5 System entropies

We can regard the action of a channel as "transferring" the probabilistic information-carrying message X into the output Y by following the conditional probability law $P_{Y|X}(y_j|x_i)$. Both the input and output ends of the channel are thus uncertain in nature: we know neither exactly which input symbol will be selected nor which output letter will be certainly seen at the output (rather, only probabilistic characterizations of various input–output events in terms of $P_X(\cdot)$ and $P_{Y|X}(\cdot|\cdot)$ are available). One immediate question to ask is: how much

[1] Note that, whereas the source is assumed to be given to us and therefore cannot be modified, we can freely choose the channel input probabilities by properly designing the channel encoder (see Figure 6.2).

"aggregate" information, or amount of uncertainty, is contained in the overall channel system? From Chapter 5, we know that the average amount of uncertainty of the input is quantified by the entropy as

$$H(X) = \sum_{i=1}^{s} P_X(x_i) \log_2 \left(\frac{1}{P_X(x_i)} \right). \tag{6.41}$$

We have shown that $H(X) \geq 0$, and $H(X) = 0$ if the input is certain; also, $H(X)$ is maximized when all x_i are equally likely.[2] We can also likewise define the entropy of the output as

$$H(Y) = \sum_{j=1}^{t} P_Y(y_j) \log_2 \left(\frac{1}{P_Y(y_j)} \right), \tag{6.42}$$

which, as expected, measures the uncertainty of the channel output. If we look at both the input and the output, the probability of the event that $X = x_i$ and $Y = y_j$ simultaneously occur is given by the joint probability $P_{X,Y}(x_i, y_j)$ (see (6.4)). Analogous to the entropy of X (or Y), we have the following definition of the entropy when both X and Y are simultaneously taken into account.

Definition 6.2 The *joint entropy* of X and Y, defined as

$$H(X,Y) \triangleq \sum_{i=1}^{s} \sum_{j=1}^{t} P_{X,Y}(x_i, y_j) \log_2 \left(\frac{1}{P_{X,Y}(x_i, y_j)} \right), \tag{6.43}$$

measures the total amount of uncertainty contained in the channel input and output, hence the overall channel system.

One might immediately ask about the relation between $H(X,Y)$ and the individual entropies, in particular whether $H(X,Y)$ just equals the sum of $H(X)$ and $H(Y)$. This is in general not true, unless X and Y are *statistically independent*, meaning that what comes out does not depend on what goes in. More precisely, independence among X and Y is characterized by [BT02]

$$P_{X,Y}(x_i, y_j) = P_X(x_i) P_Y(y_j). \tag{6.44}$$

Based on (6.43) and (6.44), we have the following proposition.

Proposition 6.3 *If X and Y are statistically independent, then*

$$H(X,Y) = H(X) + H(Y). \tag{6.45}$$

[2] As noted in Lemma 5.11, $H(X) \leq \log_2 s$ bits, with equality if $P_X(x_i) = 1/s$ for all i.

Proof With (6.44), we have

$$H(X,Y) = \sum_{i=1}^{s}\sum_{j=1}^{t} P_X(x_i)P_Y(y_j)\log_2\left(\frac{1}{P_X(x_i)P_Y(y_j)}\right) \tag{6.46}$$

$$= \sum_{i=1}^{s}\sum_{j=1}^{t} P_X(x_i)P_Y(y_j)\left(\log_2\left(\frac{1}{P_X(x_i)}\right) + \log_2\left(\frac{1}{P_Y(y_j)}\right)\right) \tag{6.47}$$

$$= \sum_{i=1}^{s}\sum_{j=1}^{t} P_X(x_i)P_Y(y_j)\log_2\left(\frac{1}{P_X(x_i)}\right) + \sum_{i=1}^{s}\sum_{j=1}^{t} P_X(x_i)P_Y(y_j)\log_2\left(\frac{1}{P_Y(y_j)}\right) \tag{6.48}$$

$$= \sum_{j=1}^{t} P_Y(y_j)\underbrace{\sum_{i=1}^{s} P_X(x_i)\log_2\left(\frac{1}{P_X(x_i)}\right)}_{H(X)} + \sum_{i=1}^{s} P_X(x_i)\underbrace{\sum_{j=1}^{t} P_Y(y_j)\log_2\left(\frac{1}{P_Y(y_j)}\right)}_{H(Y)} \tag{6.49}$$

$$= H(X) + H(Y), \tag{6.50}$$

where the last equality follows since

$$\sum_{j=1}^{t} P_Y(y_j) = \sum_{i=1}^{s} P_X(x_i) = 1. \tag{6.51}$$

□

In Proposition 6.4 we derive the relation that links the joint entropy $H(X,Y)$ with the individual $H(X)$ (or $H(Y)$) when X and Y are dependent, which is typically true since the channel output depends at least partly on the channel input (otherwise no information can be conveyed through the channel).

Proposition 6.4 (Chain rule) *The following result holds:*

$$H(X,Y) = H(X) + H(Y|X), \tag{6.52}$$

where

$$H(Y|X) \triangleq \sum_{i=1}^{s}\sum_{j=1}^{t} P_{X,Y}(x_i,y_j)\log_2\left(\frac{1}{P_{Y|X}(y_j|x_i)}\right) \tag{6.53}$$

is the conditional entropy *associated with Y given X.*

Proof By means of the relation $P_{X,Y}(x_i,y_j) = P_{Y|X}(y_j|x_i)P_X(x_i)$ (see (6.4)) it follows that

$$H(X,Y) = \sum_{i=1}^{s}\sum_{j=1}^{t} P_{X,Y}(x_i,y_j)\log_2\left(\frac{1}{P_{Y|X}(y_j|x_i)P_X(x_i)}\right) \tag{6.54}$$

$$= \sum_{i=1}^{s}\sum_{j=1}^{t} P_{X,Y}(x_i,y_j)\log_2\left(\frac{1}{P_X(x_i)}\right) + \sum_{i=1}^{s}\sum_{j=1}^{t} P_{X,Y}(x_i,y_j)\log_2\left(\frac{1}{P_{Y|X}(y_j|x_i)}\right) \tag{6.55}$$

$$= \sum_{i=1}^{s}\log_2\left(\frac{1}{P_X(x_i)}\right)\underbrace{\sum_{j=1}^{t} P_{X,Y}(x_i,y_j)}_{P_X(x_i)} + \sum_{i=1}^{s}\sum_{j=1}^{t} P_{X,Y}(x_i,y_j)\log_2\left(\frac{1}{P_{Y|X}(y_j|x_i)}\right) \tag{6.56}$$

$$= \sum_{i=1}^{s} P_X(x_i)\log_2\left(\frac{1}{P_X(x_i)}\right) + \sum_{i=1}^{s}\sum_{j=1}^{t} P_{X,Y}(x_i,y_j)\log_2\left(\frac{1}{P_{Y|X}(y_j|x_i)}\right) \tag{6.57}$$

$$= H(X) + H(Y|X). \tag{6.58}$$

□

The joint entropy $H(X,Y)$ is thus the sum of the input entropy $H(X)$ and the conditional entropy $H(Y|X)$, which measures the uncertainty remaining in Y, given that X is known. Note that if X and Y are independent, i.e. one can infer nothing about Y even if X is already known, we have $H(Y|X) = H(Y)$ and Proposition 6.4 reduces to Proposition 6.3.

Another interpretation of $H(Y|X)$ is that it represents how much must be added to the input entropy to obtain the joint entropy; in this regard, $H(Y|X)$ is called the *equivocation* of the channel. We can again use (6.4) to rewrite $H(Y|X)$ as

$$H(Y|X) = \sum_{i=1}^{s}\sum_{j=1}^{t} P_{Y|X}(y_j|x_i)P_X(x_i)\log_2\left(\frac{1}{P_{Y|X}(y_j|x_i)}\right) \tag{6.59}$$

$$= \sum_{i=1}^{s} P_X(x_i)H(Y|x_i), \tag{6.60}$$

where

$$H(Y|x_i) \triangleq \sum_{j=1}^{t} P_{Y|X}(y_j|x_i)\log_2\left(\frac{1}{P_{Y|X}(y_j|x_i)}\right) \tag{6.61}$$

is the conditional entropy of Y given a particular $X = x_i$. Finally, we remark that, starting from the alternative expression for $P_{X,Y}(x_i, y_j)$ given in (6.8), $H(X,Y)$ can be accordingly expressed as

$$H(X,Y) = H(Y) + H(X|Y). \tag{6.62}$$

Exercise 6.5 *Let (X,Y) have the joint distribution given in Table 6.1. Compute $H(X)$, $H(X,Y)$, and $H(X|Y)$.* ◇

Table 6.1 *A joint distribution of (X,Y)*

		X			
		1	2	3	4
Y	1	$\frac{1}{8}$	$\frac{1}{16}$	$\frac{1}{32}$	$\frac{1}{32}$
	2	$\frac{1}{16}$	$\frac{1}{8}$	$\frac{1}{32}$	$\frac{1}{32}$
	3	$\frac{1}{16}$	$\frac{1}{16}$	$\frac{1}{16}$	$\frac{1}{16}$
	4	$\frac{1}{4}$	0	0	0

Exercise 6.6 *Verify that $H(Y|X) = H(Y)$ if X and Y are independent.* ◇

6.6 Mutual information

Consider again the transmission system shown in Figure 6.3. We wish to determine how much information about the input can be gained based on some particular received output letter $Y = y_j$; this is the first step toward quantifying the amount of information that can get through the channel.

At the transmit side, the probability that the ith input symbol x_i occurs is $P_X(x_i)$, which is called the *a priori*[3] *probability* of x_i. Upon receiving $Y = y_j$, one can try to infer which symbol probably has been sent based on the information carried by y_j. In particular, given y_j is received, the probability that x_i has been sent is given by the backward conditional probability $P_{X|Y}(x_i|y_j)$, which is commonly termed the *a posteriori*[4] *probability* of x_i. The change of probability (from a priori to a posteriori) is closely related to how much information one can learn about x_i from the reception of y_j. Specifically, the difference between the uncertainty before and after receiving y_j measures the

[3] From the Latin, meaning "from what comes first" or "before."

[4] From the Latin, meaning "from what comes after" or "afterwards."

gain in information due to the reception of y_j. Such an information gain is called the *mutual information* and is naturally defined to be

$$\underbrace{I(x_i;y_j)}_{\substack{\text{information gain}\\ \text{or uncertainty loss}\\ \text{after receiving } y_j}} \triangleq \underbrace{\log_2\left(\frac{1}{P_X(x_i)}\right)}_{\substack{\text{uncertainty}\\ \text{before receiving } y_j}} - \underbrace{\log_2\left(\frac{1}{P_{X|Y}(x_i|y_j)}\right)}_{\substack{\text{uncertainty}\\ \text{after receiving } y_j}} \tag{6.63}$$

$$= \log_2\left(\frac{P_{X|Y}(x_i|y_j)}{P_X(x_i)}\right). \tag{6.64}$$

Note that if the two events $X = x_i$ and $Y = y_j$ are independent, thereby

$$P_{X|Y}(x_i|y_j) = P_X(x_i), \tag{6.65}$$

we have $I(x_i;y_j) = 0$, i.e. no information about x_i is gained once y_j is received.

For the noiseless channel, thus $y_j = x_i$, we have $P_{X|Y}(x_i|y_j) = 1$ since, based on what is received, we are completely certain about which input symbol has been sent. In this case, the mutual information attains the maximum value $\log_2(1/P_X(x_i))$; this means that all information about x_i is conveyed without any loss over the channel.

Since

$$P_{X|Y}(x_i|y_j)P_Y(y_j) = P_{X,Y}(x_i,y_j) = P_{Y|X}(y_j|x_i)P_X(x_i), \tag{6.66}$$

we have

$$I(x_i;y_j) = \log_2\left(\frac{P_{X,Y}(x_i,y_j)}{P_X(x_i)P_Y(y_j)}\right) = I(y_j;x_i). \tag{6.67}$$

Hence, we see that x_i provides the same amount of information about y_j as y_j does about x_i. This is why $I(x_i;y_j)$ has been coined "*mutual* information."

We have now characterized the mutual information with respect to a particular input–output event. Owing to the random nature of the source and channel output, the mutual information should be averaged with respect to both the input and output in order to account for the true statistical behavior of the channel. This motivates the following definition.

Definition 6.7 The *system mutual information*, or *average mutual information*, is defined as

$$I(X;Y) \triangleq \sum_{i=1}^{s}\sum_{j=1}^{t} P_{X,Y}(x_i,y_j)I(x_i;y_j) \tag{6.68}$$

$$= \sum_{i=1}^{s}\sum_{j=1}^{t} P_{X,Y}(x_i,y_j)\log_2\left(\frac{P_{X,Y}(x_i,y_j)}{P_X(x_i)P_Y(y_j)}\right). \tag{6.69}$$

The average mutual information $I(X;Y)$ has the following properties (the proofs are left as exercises).

Lemma 6.8 *The system mutual information has the following properties:*

(1) $I(X;Y) \geq 0$*;*
(2) $I(X;Y) = 0$ *if, and only if,* X *and* Y *are independent;*
(3) $I(X;Y) = I(Y;X)$.

Exercise 6.9 *Prove Lemma 6.8.*

Hint: For the first and second property use the IT Inequality (Lemma 5.10). You may proceed similarly to the proof of Lemma 5.11. The third property can be proven based on the formula of the mutual information, i.e. (6.69). ◊

Starting from (6.68), we have

$$I(X;Y) = \sum_{i=1}^{s}\sum_{j=1}^{t} P_{X,Y}(x_i,y_j)I(x_i;y_j) \tag{6.70}$$

$$= \sum_{i=1}^{s}\sum_{j=1}^{t} P_{X|Y}(x_i|y_j)P_Y(y_j)I(x_i;y_j) \tag{6.71}$$

$$= \sum_{j=1}^{t} P_Y(y_j)I(X;y_j), \tag{6.72}$$

where

$$I(X;y_j) \triangleq \sum_{i=1}^{s} P_{X|Y}(x_i|y_j)I(x_i;y_j) \tag{6.73}$$

measures the information about the entire input X provided by the reception of the particular y_j. In an analogous way we can obtain

$$I(X;Y) = \sum_{i=1}^{s}\sum_{j=1}^{t} P_{X,Y}(x_i,y_j)I(x_i;y_j) \tag{6.74}$$

$$= \sum_{i=1}^{s}\sum_{j=1}^{t} P_{Y|X}(y_j|x_i)P_X(x_i)I(x_i;y_j) \tag{6.75}$$

$$= \sum_{i=1}^{s} P_X(x_i)I(x_i;Y), \tag{6.76}$$

where

$$I(x_i;Y) \triangleq \sum_{j=1}^{t} P_{Y|X}(y_j|x_i)I(x_i;y_j) \tag{6.77}$$

represents the information about the output Y given that we know the input letter x_i is sent.

Let us end this section by specifying the relation between the average mutual information $I(X;Y)$ and various information quantities introduced thus far, e.g. input entropy $H(X)$, output entropy $H(Y)$, joint entropy $H(X,Y)$, and conditional entropies $H(X|Y)$ and $H(Y|X)$. To proceed, let us use (6.69) to express $I(X;Y)$ as

$$
\begin{aligned}
I(X;Y) &= \sum_{i=1}^{s}\sum_{j=1}^{t} P_{X,Y}(x_i,y_j)\log_2\left(\frac{P_{X,Y}(x_i,y_j)}{P_X(x_i)P_Y(y_j)}\right) && (6.78)\\
&= \sum_{i=1}^{s}\sum_{j=1}^{t} P_{X,Y}(x_i,y_j)\Big(\log_2 P_{X,Y}(x_i,y_j) - \log_2 P_X(x_i) \\
&\qquad - \log_2 P_Y(y_j)\Big) && (6.79)\\
&= -\sum_{i=1}^{s}\sum_{j=1}^{t} P_{X,Y}(x_i,y_j)\log_2\left(\frac{1}{P_{X,Y}(x_i,y_j)}\right) \\
&\quad + \sum_{i=1}^{s} P_X(x_i)\log_2\left(\frac{1}{P_X(x_i)}\right) + \sum_{j=1}^{t} P_Y(y_j)\log_2\left(\frac{1}{P_Y(y_j)}\right) && (6.80)\\
&= H(X) + H(Y) - H(X,Y) \geq 0. && (6.81)
\end{aligned}
$$

Since

$$
\begin{aligned}
H(X,Y) &= H(X) + H(Y|X) && (6.82)\\
&= H(Y) + H(X|Y), && (6.83)
\end{aligned}
$$

we also have

$$
\begin{aligned}
I(X;Y) &= H(X) - H(X|Y) && (6.84)\\
&= H(Y) - H(Y|X). && (6.85)
\end{aligned}
$$

We have the following corollary (the proof is left as an exercise).

Corollary 6.10 (Conditioning reduces entropy) *The following inequalities hold:*

$$
\begin{aligned}
&(a) && 0 \leq H(X|Y) \leq H(X), && (6.86)\\
& && 0 \leq H(Y|X) \leq H(Y); && (6.87)\\
&(b) && H(X,Y) \leq H(X) + H(Y). && (6.88)
\end{aligned}
$$

Part (a) of Corollary 6.10 asserts that conditioning cannot increase entropy, whereas Part (b) shows that the joint entropy $H(X,Y)$ is maximized when X and Y are independent.

Exercise 6.11 *Prove Corollary 6.10.*
Hint: Use known properties of $I(X;Y)$ and $H(X)$. ◊

To summarize: the average mutual information is given by

$$I(X;Y) = \begin{cases} H(X)+H(Y)-H(X,Y), \\ H(X)-H(X|Y), \\ H(Y)-H(Y|X); \end{cases} \tag{6.89}$$

the equivocation is given by

$$H(X|Y) = H(X) - I(X;Y), \tag{6.90}$$

$$H(Y|X) = H(Y) - I(X;Y); \tag{6.91}$$

the joint entropy is given by

$$H(X,Y) = \begin{cases} H(X)+H(Y)-I(X;Y), \\ H(X)+H(Y|X), \\ H(Y)+H(X|Y). \end{cases} \tag{6.92}$$

A schematic description of the relations between various information quantities is given by the Venn diagram in Figure 6.5.

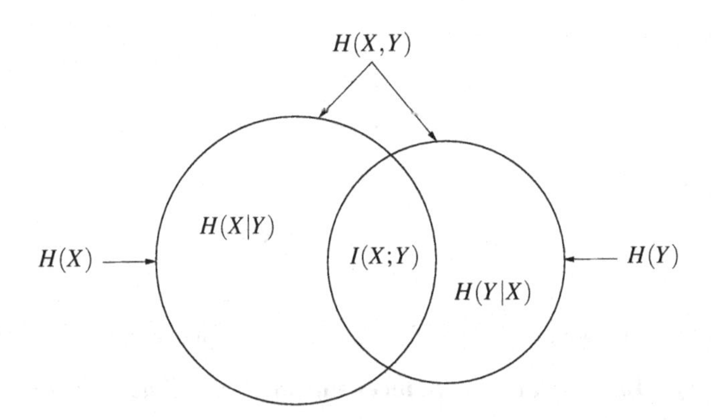

Figure 6.5 Relation between entropy, conditional entropy, and mutual information.

6.7 Definition of channel capacity

Given the conditional probabilities $P_{Y|X}(y_j|x_i)$, which define a channel, what is the maximum amount of information we can send through the channel? This is the main question attacked in the rest of this chapter.

The mutual information connects the two ends of the channel together. It is defined by (6.84) as

$$I(X;Y) = H(X) - H(X|Y), \tag{6.93}$$

where the entropy $H(X)$ is the uncertainty of the channel input *before* the reception of Y, and $H(X|Y)$ is the uncertainty that remains *after* the reception of Y. Thus $I(X;Y)$ is the change in the uncertainty. An alternative expression for $I(X;Y)$ is

$$I(X;Y) = \sum_{i=1}^{s}\sum_{j=1}^{t} P_X(x_i)P_{Y|X}(y_j|x_i)\log_2\left(\frac{P_{Y|X}(y_j|x_i)}{\sum_{i'=1}^{s} P_X(x_{i'})P_{Y|X}(y_j|x_{i'})}\right). \tag{6.94}$$

This formula involves the input symbol frequencies $P_X(x_i)$; in particular, for a given channel law $P_{Y|X}(y_j|x_i)$, $I(X;Y)$ depends completely on $P_X(x_i)$. We saw in the example of a binary symmetric channel (Section 6.4) how a poor match of $P_X(x_i)$ to the channel can ruin a channel. Indeed, we know that if the probability of one symbol is $P_X(x_i) = 1$, then all the others must be zero and the constant signal contains no information.

How can we best choose the $P_X(x_i)$ to get the most through the channel, and what is that amount?

Definition 6.12 (Capacity) For a given channel, the *channel capacity*, denoted by C, is defined to be the maximal achievable system mutual information $I(X;Y)$ among all possible input distributions $P_X(\cdot)$:

$$\mathrm{C} \triangleq \max_{P_X(\cdot)} I(X;Y). \tag{6.95}$$

Finding a closed-form solution to the channel capacity is in general difficult, except for some simple channels, e.g. the binary symmetric channel defined in Section 6.4 (see also Section 6.8 below).

We would like to point out that even though this definition of capacity is intuitively quite pleasing, at this stage it is a mere mathematical quantity, i.e. a number that is the result of a maximization problem. However, we will see in Section 6.11 that it really is the capacity of a channel in the sense that it is only possible to transmit signals reliably (i.e. with very small error probability) through the channel as long as the transmission rate is below the capacity.

6.8 Capacity of the binary symmetric channel

Consider again the binary symmetric channel (BSC), with the probability of transmission error equal to ε, as depicted in Figure 6.6.

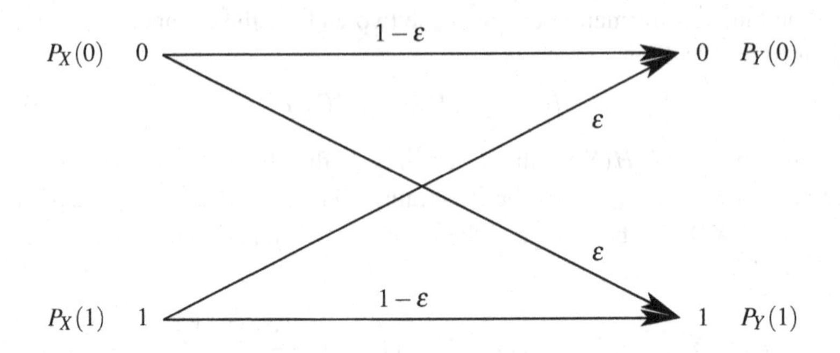

Figure 6.6 Binary symmetric channel (BSC).

The channel matrix is given by

$$\begin{pmatrix} 1-\varepsilon & \varepsilon \\ \varepsilon & 1-\varepsilon \end{pmatrix}. \tag{6.96}$$

Exercise 6.13 *For the BSC, show that*

$$H(Y|X=0) = H(Y|X=1) = H_{\mathrm{b}}(\varepsilon), \tag{6.97}$$

where $H_{\mathrm{b}}(\cdot)$ is the binary entropy function defined in (5.24). ◊

We start from the definition of mutual information (6.85) to obtain the following set of relations:

$$\begin{aligned} I(X;Y) &= H(Y) - H(Y|X) && (6.98)\\ &= H(Y) - \sum_{x\in\mathcal{X}} P_X(x)H(Y|X=x) && (6.99)\\ &= H(Y) - \big(P_X(0)H(Y|X=0) + P_X(1)H(Y|X=1)\big) && (6.100)\\ &= H(Y) - H_{\mathrm{b}}(\varepsilon) && (6.101)\\ &\le 1 - H_{\mathrm{b}}(\varepsilon) \text{ bits}, && (6.102) \end{aligned}$$

where (6.102) follows since Y is a binary random variable (see Lemma 5.11). Since equality in (6.102) is attained if Y is uniform, which will hold if input X is uniform, we conclude that the capacity of the BSC is given by

$$\mathrm{C} = 1 - H_{\mathrm{b}}(\varepsilon) \text{ bits}, \tag{6.103}$$

and that the achieving input distribution is $P_X(0) = P_X(1) = 1/2$. Alternatively, we can find the capacity of a BSC by starting from $P_X(0) = \delta = 1 - P_X(1)$ and expressing $I(X;Y)$ as

$$I(X;Y) = H(Y) - H(Y|X) \tag{6.104}$$

$$\begin{aligned} &= -\big(\delta(1-\varepsilon)+(1-\delta)\varepsilon\big)\log_2\big(\delta(1-\varepsilon)+(1-\delta)\varepsilon\big) \\ &\quad -\big(\delta\varepsilon+(1-\delta)(1-\varepsilon)\big)\log_2\big(\delta\varepsilon+(1-\delta)(1-\varepsilon)\big) \\ &\quad +(1-\varepsilon)\log_2(1-\varepsilon)+\varepsilon\log_2\varepsilon. \end{aligned} \tag{6.105}$$

If we now maximize the above quantity over $\delta \in [0,1]$, we find that the optimal δ is $\delta = 1/2$, which immediately yields (6.103).

Figure 6.7 depicts the mutual information in (6.105) versus δ with respect to three different choices of the error probability ε. As can be seen from the figure, the peak value of each curve is indeed attained by $\delta = 1/2$.

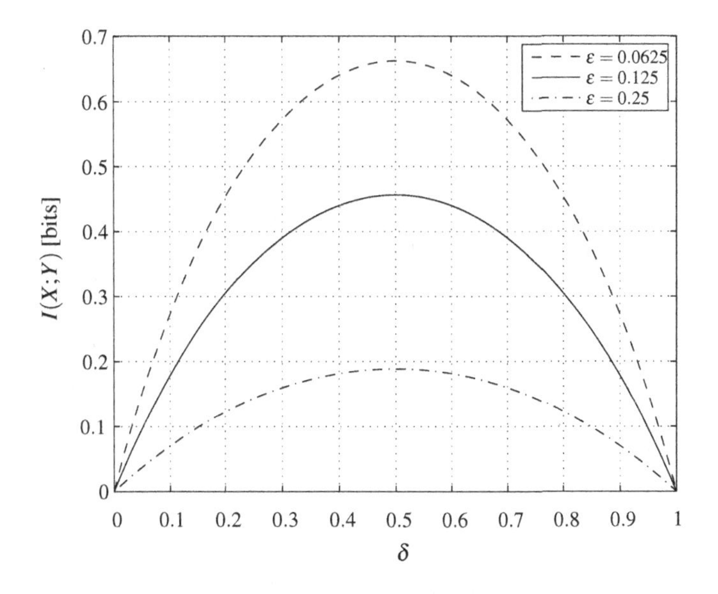

Figure 6.7 Mutual information over the binary symmetric channel (BSC): (6.105) as a function of δ, for various values of ε.

Figure 6.8 plots the capacity C in (6.103) versus the cross-over probability ε. We see from the figure that C attains the maximum value 1 bit when $\varepsilon = 0$ or $\varepsilon = 1$, and attains the minimal value 0 when $\varepsilon = 1/2$.

When $\varepsilon = 0$, it is easy to see that $C = 1$ bit is the maximum rate at which information can be communicated through the channel reliably. This can be achieved simply by transmitting uncoded bits through the channel, and no decoding is necessary because the bits are received unchanged. When $\varepsilon = 1$ the

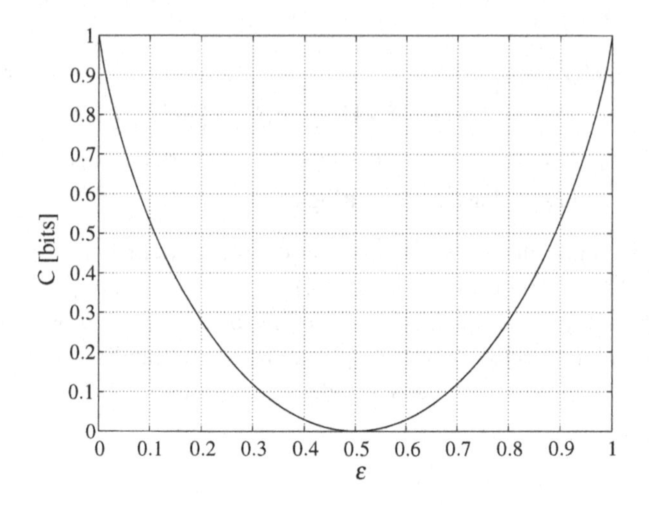

Figure 6.8 Capacity of the binary symmetric channel (BSC).

same can be achieved with the additional decoding step which complements all the received bits. By doing so, the bits transmitted through the channel can be recovered without error. Thus from a communications point of view, for binary channels, a channel which never makes an error and a channel which always makes an error are equally good.

When $\varepsilon = 1/2$, the channel output is independent of the channel input. Therefore, no information can possibly be communicated through the channel.

6.9 Uniformly dispersive channel

Recall that the channel transition matrix for the BSC is

$$\begin{pmatrix} 1-\varepsilon & \varepsilon \\ \varepsilon & 1-\varepsilon \end{pmatrix}, \tag{6.106}$$

in which the second row is a permutation of the first row. In fact, the BSC belongs to the class of *uniformly dispersive channels*.

Definition 6.14 A channel is said to be *uniformly dispersive* if the set

$$\mathcal{A}(x) \triangleq \{P_{Y|X}(y_1|x), \ldots, P_{Y|X}(y_t|x)\} \tag{6.107}$$

is identical for each input symbol x. Hence a uniformly dispersive channel has a channel matrix

$$\begin{pmatrix} P_{11} & P_{12} & \cdots & P_{1t} \\ P_{21} & P_{22} & \cdots & P_{2t} \\ \vdots & \vdots & \ddots & \vdots \\ P_{s1} & P_{s2} & \cdots & P_{st} \end{pmatrix} \tag{6.108}$$

such that each row is a permutation of the first row.

According to the definition, for a uniformly dispersive channel the entropy of the output conditioned on a particular input alphabet x being sent, namely

$$H(Y|X=x) = \sum_{y \in \mathcal{Y}} P_{Y|X}(y|x) \log_2 \left(\frac{1}{P_{Y|X}(y|x)} \right), \tag{6.109}$$

is thus identical for all x. By means of (6.109), the mutual information between the channel input and output reads

$$\begin{aligned} I(X;Y) &= H(Y) - H(Y|X) && (6.110) \\ &= H(Y) - \sum_{x \in \mathcal{X}} P_X(x) H(Y|X=x) && (6.111) \\ &= H(Y) - H(Y|X=x) \underbrace{\sum_{x \in \mathcal{X}} P_X(x)}_{=1} && (6.112) \\ &= H(Y) - H(Y|X=x). && (6.113) \end{aligned}$$

Equations (6.110)–(6.113) should be reminiscent of (6.98)–(6.101). This is no surprise since the BSC is uniformly dispersive.

Recall also from (6.102) that the capacity of the BSC is attained with a uniform output, which can be achieved by a uniform input. At a first glance one might expect that a similar argument can be directly applied to the uniformly dispersive case to find the capacity of an arbitrary uniformly dispersive channel. However, for a general uniformly dispersive channel, the uniform input does not necessarily result in the uniform output, and neither can the capacity necessarily be achieved with the uniform output. For example, consider the binary erasure channel (BEC) depicted in Figure 6.9. In this channel, the input alphabet is $\mathcal{X} = \{0, 1\}$, while the output alphabet is $\mathcal{Y} = \{0, 1, ?\}$. With probability γ, the erasure symbol ? is produced at the output, which means that the input bit is lost; otherwise the input bit is reproduced at the output without error. The parameter γ is thus called the *erasure probability*. The BEC has the channel transition matrix

$$\begin{pmatrix} 1-\gamma & \gamma & 0 \\ 0 & \gamma & 1-\gamma \end{pmatrix} \tag{6.114}$$

and is thus, by definition, uniformly dispersive (the second row is a permutation of the first row). However, with the input distribution $P_X(0) = P_X(1) = 1/2$, the output is, in general, not uniform ($P_Y(?) = \gamma$ and $P_Y(0) = P_Y(1) = (1-\gamma)/2$). Despite this, the uniform input remains as the capacity-achieving input distribution for the BEC.

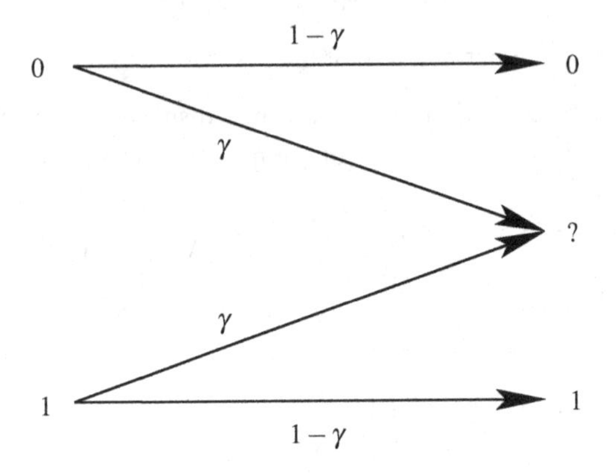

Figure 6.9 Binary erasure channel (BEC).

Exercise 6.15 *Based on (6.113), show that the capacity of the BEC is*

$$C = 1 - \gamma \text{ bits}, \tag{6.115}$$

which is attained with $P_X(0) = P_X(1) = 1/2$. *The result is intuitively reasonable: since a proportion* γ *of the bits are lost in the channel, we can recover (at most) a proportion* $(1-\gamma)$ *of the bits, and hence the capacity is* $(1-\gamma)$. ◊

6.10 Characterization of the capacity-achieving input distribution

Even though it is, in general, difficult to find the closed-form capacity formula and the associated capacity-achieving input distribution, it is nonetheless possible to specify some underlying properties of the optimal $P_X(\cdot)$. The following theorem, which is stated without proof, provides one such characterization in terms of $I(x;Y)$, i.e. the information gain about Y given that $X = x$ is sent (see (6.77)).

Theorem 6.16 (Karush–Kuhn–Tucker (KKT) conditions) *An input distribution $P_X(\cdot)$ achieves the channel capacity* C *if, and only if,*

$$I(x;Y)\begin{cases} = \mathsf{C} & \text{for all } x \text{ with } P_X(x) > 0; \\ \le \mathsf{C} & \text{for all } x \text{ with } P_X(x) = 0. \end{cases} \tag{6.116}$$

Remark 6.17 The KKT conditions were originally named after Harold W. Kuhn and Albert W. Tucker, who first published the conditions in 1951 [KT51]. Later, however, it was discovered that the necessary conditions for this problem had already been stated by William Karush in his master's thesis [Kar39].

The assertion of Theorem 6.16 is rather intuitive: if $P_X(\cdot)$ is the capacity-achieving input distribution and $P_X(x) > 0$, i.e. the particular letter x will be used with a nonvanishing probability to convey information over the channel, then the contribution of the mutual information due to this x must attain the capacity; otherwise there will exist another $P_{X'}(\cdot)$ capable of achieving the capacity by just disregarding this x (thus, $P_{X'}(x) = 0$) and using more often input letters other than x.

Theorem 6.16 can also be exploited for finding the capacity of some channels. Consider again the BSC case; the capacity should satisfy one of the following three cases:

$$\mathsf{C} = I(0;Y) = I(1;Y) \quad \text{for } P_X(0) > 0 \text{ and } P_X(1) > 0 \tag{6.117}$$

or

$$\mathsf{C} = I(0;Y) \ge I(1;Y) \quad \text{for } P_X(0) = 1 \text{ and } P_X(1) = 0 \tag{6.118}$$

or

$$\mathsf{C} = I(1;Y) \ge I(0;Y) \quad \text{for } P_X(0) = 0 \text{ and } P_X(1) = 1. \tag{6.119}$$

Since (6.118) and (6.119) only yield uninteresting zero capacity, it remains to verify whether or not (6.117) can give a positive capacity. By rearrangement (6.117) implies

$$\begin{aligned}
\mathsf{C} &= I(0;Y) && (6.120)\\
&= \sum_{y=0}^{1} P_{Y|X}(y|0)\log_2\frac{P_{Y|X}(y|0)}{P_Y(y)} && (6.121)\\
&= -\sum_{y=0}^{1} P_{Y|X}(y|0)\log_2 P_Y(y) + \sum_{y=0}^{1} P_{Y|X}(y|0)\log_2 P_{Y|X}(y|0) && (6.122)\\
&= -(1-\varepsilon)\log_2 P_Y(0) - \varepsilon\log_2 P_Y(1) - H_\mathrm{b}(\varepsilon) && (6.123)
\end{aligned}$$

and

$$\begin{aligned} \mathsf{C} &= I(1;Y) && (6.124) \\ &= \sum_{y=0}^{1} P_{Y|X}(y|1) \log_2 \frac{P_{Y|X}(y|1)}{P_Y(y)} && (6.125) \\ &= -\sum_{y=0}^{1} P_{Y|X}(y|1) \log_2 P_Y(y) + \sum_{y=0}^{1} P_{Y|X}(y|1) \log_2 P_{Y|X}(y|1) && (6.126) \\ &= -\varepsilon \log_2 P_Y(0) - (1-\varepsilon) \log_2 P_Y(1) - H_b(\varepsilon), && (6.127) \end{aligned}$$

which yields

$$\begin{aligned} &-(1-\varepsilon) \cdot \log_2 P_Y(0) - \varepsilon \cdot \log_2 P_Y(1) - H_b(\varepsilon) \\ &\quad = -\varepsilon \cdot \log_2 P_Y(0) - (1-\varepsilon) \cdot \log_2 P_Y(1) - H_b(\varepsilon). \qquad (6.128) \end{aligned}$$

This can only be satisfied if $P_Y(0) = P_Y(1)$ $(= 1/2)$. Thus

$$\begin{aligned} \mathsf{C} &= -\varepsilon \cdot \log_2 \left(\frac{1}{2}\right) - (1-\varepsilon) \cdot \log_2 \left(\frac{1}{2}\right) - H_b(\varepsilon) && (6.129) \\ &= 1 - H_b(\varepsilon) \text{ bits.} && (6.130) \end{aligned}$$

Exercise 6.18 *Repeat the above arguments to derive the capacity of the BEC.* ◊

6.11 Shannon's Channel Coding Theorem

The channel capacity measures the amount of information that can be carried over the channel; in fact, it characterizes the maximal amount of transmission rate for reliable communication. Prior to the mid 1940s people believed that transmitted data subject to noise corruption can never be perfectly recovered unless the transmission rate approaches zero [Gal01]. Shannon's landmark work [Sha48] in 1948 disproved this thinking and established the well known *Channel Coding Theorem*: as long as the transmission rate in the same units as the channel capacity, e.g. information bits per channel use, is below (but can be arbitrarily close to) the channel capacity, the error can be made *smaller than any given number* (which we term *arbitrarily small*) by some properly designed coding scheme.

In what follows are some definitions that are required to state the theorem formally; detailed mathematical proofs can be found in [CT06] and [Gal68].

Definition 6.19 An (M, n) *coding scheme* for the channel $(\mathcal{X}, P_{Y|X}(y|x), \mathcal{Y})$ consists of the following.

(1) A message set $\{1,2,\ldots,\mathsf{M}\}$.

(2) An encoding function $\phi\colon \{1,2,\ldots,\mathsf{M}\} \to \mathcal{X}^n$, which is a rule that associates message m with a channel input sequence of length n, called the mth *codeword* and denoted by $\mathbf{x}^n(m)$. The set of all codewords

$$\{\mathbf{x}^n(1),\mathbf{x}^n(2),\ldots,\mathbf{x}^n(\mathsf{M})\}$$

is called the *codebook* (or simply the *code*).

(3) A decoding function $\psi\colon \mathcal{Y}^n \to \{1,2,\ldots,\mathsf{M}\}$, which is a deterministic rule that assigns a guess to each possible received vector.

Definition 6.20 (Rate) The *rate* R of an (M,n) coding scheme is defined to be

$$\mathsf{R} \triangleq \frac{\log_2 \mathsf{M}}{n} \quad \text{bits per transmission.} \tag{6.131}$$

In (6.131), $\log_2 \mathsf{M}$ describes the number of digits needed to write the numbers $0,\ldots,\mathsf{M}-1$ in binary form. For example, for $\mathsf{M}=8$ we need three binary digits (or bits): $000,\ldots,111$. The denominator n tells how many times the channel is used for the total transmission of a codeword (recall that n is the codeword length). Hence the rate describes how many bits are transmitted on average in each channel use.

Definition 6.21 Let

$$\lambda_m \triangleq \Pr[\psi(\mathbf{Y}^n) \neq m \,|\, \mathbf{X}^n = \mathbf{x}^n(m)] \tag{6.132}$$

be the conditional probability that the receiver makes a wrong guess given that the mth codeword is sent. The *average error probability* $\lambda^{(n)}$ for an (M,n) coding scheme is defined as

$$\lambda^{(n)} \triangleq \frac{1}{\mathsf{M}} \sum_{m=1}^{\mathsf{M}} \lambda_m. \tag{6.133}$$

Now we are ready for the famous *Channel Coding Theorem* due to Shannon.

Theorem 6.22 (Shannon's Channel Coding Theorem)
For a discrete-time information channel, it is possible to transmit messages with an arbitrarily small error probability (i.e. we have so-called reliable communication*), if the communication rate* R *is below the channel capacity* C. *Specifically, for every rate* $\mathrm{R} < \mathrm{C}$, *there exists a sequence of* $(2^{n\mathrm{R}}, n)$ *coding schemes*[5] *with average error probability* $\lambda^{(n)} \to 0$ *as* $n \to \infty$.
Conversely, any sequence of $(2^{n\mathrm{R}}, n)$ *coding schemes with* $\lambda^{(n)} \to 0$ *must have* $\mathrm{R} \leq \mathrm{C}$. *Hence, any attempt of transmitting at a rate larger than capacity will for sure fail in the sense that the average error probability is strictly larger than zero.*

Take the BSC for example. If the cross-over probability is $\varepsilon = 0.1$, the resulting capacity is $\mathrm{C} = 0.531$ bits per channel use. Hence reliable communication is only possible for coding schemes with a rate smaller than 0.531 bits per channel use.

Although the theorem shows that there exist good coding schemes with arbitrarily small error probability for long blocklength n, it does not provide a way of constructing the best coding schemes. Actually, the only knowledge we can infer from the theorem is perhaps "a good code favors a large blocklength." Ever since Shannon's original findings, researchers have tried to develop practical coding schemes that are easy to encode and decode; the Hamming code we discussed in Chapter 3 is the simplest of a class of algebraic error-correcting codes that can correct one error in a block of bits. Many other techniques have also been proposed to construct error-correcting codes, among which the *turbo code* – to be discussed in Chapter 7 – has come close to achieving the so-called *Shannon limit* for channels contaminated by Gaussian noise.

6.12 Some historical background

In his landmark paper [Sha48], Shannon only used H, R, and C to denote entropy, rate, and capacity, respectively. The first to use I for information were

[5] In Theorem 6.22, $2^{n\mathrm{R}}$ is a convenient expression for the code size and should be understood as either the smallest integer no less than its value or the largest integer no greater than its value. Researchers tend to drop the ceiling or flooring function applying to it, because the ratio of $2^{n\mathrm{R}}$, against the integer it is understood to be, will be very close to unity as n is large. Since the theorem actually deals with very large codeword lengths n (note that $\lambda^{(n)}$ approaches zero only when n is very large), the slack use of $2^{n\mathrm{R}}$ as an integer is somehow justified in concept.

Philip M. Woodward and Ian L. Davies in [WD52]. This paper is a very good read and gives an astoundingly clear overview of the fundamentals of information theory only four years after the theory had been established by Shannon. The authors give a slightly different interpretation of Shannon's theory and redevelop it using two additivity axioms. However, they did not yet use the name "mutual information." The name only starts to appear between 1954 and 1956. In 1954, Mark Pinsker published a paper in Russian [Pin54] with the title "Mutual information between a pair of stationary Gaussian random processes." However, depending on the translation, the title also might read "The quantity of information about a Gaussian random stationary process, contained in a second process connected with it in a stationary manner." Shannon certainly used the term "mutual information" in a paper about the zero-error capacity in 1956 [Sha56].

By the way, Woodward is also a main pioneer in modern radar theory. He had the insight to apply probability theory and statistics to the problem of recovering data from noisy samples. Besides this, he is a huge clock fan and made many contributions to horology; in particular, he built the world's most precise mechanical clock, the *Clock W5*, inventing a completely new mechanism.[6]

6.13 Further reading

Full discussions of the mutual information, channel capacity, and Shannon's Channel Coding Theorem in terms of probability theory can be found in many textbooks, see, e.g., [CT06] and [Gal68]. A unified discussion of the capacity results of the uniformly dispersive channel is given in [Mas96]. Further generalizations of uniformly dispersive channels are *quasi-symmetric channels*, discussed in [CA05, Chap. 4], and *T-symmetric channels*, described in [RG04]. The proof of Theorem 6.16 is closely related to the subject of constrained optimization, which is a standard technique for finding the channel capacity; see, e.g., [Gal68] and [CT06]. In addition to the Source Coding Theorem (Theorem 5.28 introduced in Chapter 5) and the Channel Coding Theorem (Theorem 6.22), Shannon's third landmark contribution is the development of the so-called *rate distortion theory*, which describes how to represent a continuous source with good fidelity using only a finite number of "representation levels"; for more details, please also refer to [CT06] and [Gal68].

[6] To see some amazing videos of this, search on http://www.youtube.com for "clock W5."

References

[BT02] Dimitri P. Bertsekas and John N. Tsitsiklis, *Introduction to Probability*. Athena Scientific, Belmont, MA, 2002.

[CA05] Po-Ning Chen and Fady Alajaji, *Lecture Notes on Information Theory,* vol. 1, Department of Electrical Engineering, National Chiao Tung University, Hsinchu, Taiwan, and Department of Mathematics & Statistics, Queen's University, Kingston, Canada, August 2005. Available: http://shannon.cm.nctu.edu.tw/it/itvol12004.pdf

[CT06] Thomas M. Cover and Joy A. Thomas, *Elements of Information Theory*, 2nd edn. John Wiley & Sons, Hoboken, NJ, 2006.

[Gal68] Robert G. Gallager, *Information Theory and Reliable Communication*. John Wiley & Sons, New York, 1968.

[Gal01] Robert G. Gallager, "Claude E. Shannon: a retrospective on his life, work, and impact," *IEEE Transactions on Information Theory*, vol. 47, no. 7, pp. 2681–2695, November 2001.

[Kar39] William Karush, "Minima of functions of several variables with inequalities as side constraints," Master's thesis, Department of Mathematics, University of Chicago, Chicago, IL, 1939.

[KT51] Harold W. Kuhn and Albert W. Tucker, "Nonlinear programming," in *Proceedings of Second Berkeley Symposium on Mathematical Statistics and Probability*, J. Neyman, ed. University of California Press, Berkeley, CA, 1951, pp. 481–492.

[Mas96] James L. Massey, *Applied Digital Information Theory I and II,* Lecture notes, Signal and Information Processing Laboratory, ETH Zurich, 1995/1996. Available: http://www.isiweb.ee.ethz.ch/archive/massey_scr/

[Pin54] Mark S. Pinsker, "Mutual information between a pair of stationary Gaussian random processes," (in Russian), *Doklady Akademii Nauk SSSR*, vol. 99, no. 2, pp. 213–216, 1954, also known as "The quantity of information about a Gaussian random stationary process, contained in a second process connected with it in a stationary manner."

[RG04] Mohammad Rezaeian and Alex Grant, "Computation of total capacity for discrete memoryless multiple-access channels," *IEEE Transactions on Information Theory*, vol. 50, no. 11, pp. 2779–2784, November 2004.

[Sha48] Claude E. Shannon, "A mathematical theory of communication," *Bell System Technical Journal*, vol. 27, pp. 379–423 and 623–656, July and October 1948. Available: http://moser.cm.nctu.edu.tw/nctu/doc/shannon1948.pdf

[Sha56] Claude E. Shannon, "The zero error capacity of a noisy channel," *IRE Transactions on Information Theory*, vol. 2, no. 3, pp. 8–19, September 1956.

[WD52] Philip M. Woodward and Ian L. Davies, "Information theory and inverse probability in telecommunication," *Proceedings of the IEE*, vol. 99, no. 58, pp. 37–44, March 1952.

7
Approaching the Shannon limit by turbo coding

7.1 Information Transmission Theorem

The reliable transmission of information-bearing signals over a noisy communication channel is at the heart of what we call communication. Information theory, founded by Claude E. Shannon in 1948 [Sha48], provides a mathematical framework for the theory of communication. It describes the fundamental limits to how efficiently one can encode information and still be able to recover it with negligible loss.

At its inception, the main role of information theory was to provide the engineering and scientific communities with a mathematical framework for the theory of communication by establishing the fundamental limits on the performance of various communication systems. Its birth was initiated with the publication of the works of Claude E. Shannon, who stated that it is possible to send information-bearing signals at a fixed code rate through a noisy communication channel with an arbitrarily small error probability as long as the code rate is below a certain fixed quantity that depends on the channel characteristics [Sha48]; he "baptized" this quantity with the name of *channel capacity* (see the discussion in Chapter 6). He further proclaimed that random sources – such as speech, music, or image signals – possess an irreducible complexity beyond which they cannot be compressed distortion-free. He called this complexity the *source entropy* (see the discussion in Chapter 5). He went on to assert that if a source has an entropy that is less than the capacity of a communication channel, then asymptotically error-free transmission of the source over the channel can be achieved. This result is usually referred to as the *Information Transmission Theorem* or the *Joint Source–Channel Coding Theorem.*

Theorem 7.1 (Information Transmission Theorem)
Consider the transmission of a source $\mathbf{U}^k = (U_1, U_2, \ldots, U_k)$ *through a channel with input* $\mathbf{X}^n = (X_1, X_2, \ldots, X_n)$ *and output* $\mathbf{Y}^n = (Y_1, Y_2, \ldots, Y_n)$ *as shown in Figure 7.1. Assume that both the source sequence* $U_1, U_2, \ldots, U_k$ *and the noise sequence* $N_1, N_2, \ldots, N_n$ *are independent and identically distributed. Then, subject to a fixed code rate* $\mathrm{R} = k/n$*, there exists a sequence of encoder–decoder pairs such that the decoding error, i.e.* $\Pr\left[\hat{\mathbf{U}}^k \neq \mathbf{U}^k\right]$*, can be made arbitrarily small (i.e. arbitrarily close to zero) by taking n sufficiently large if*

$$\frac{1}{\mathrm{T_s}} H(U) \text{ bits/second} < \frac{1}{\mathrm{T_c}} \max_{P_X} I(X;Y) \text{ bits/second}, \tag{7.1}$$

where the base-2 logarithm is adopted in the calculation of entropy and mutual information (so they are in units of bits), and $\mathrm{T_s}$ *and* $\mathrm{T_c}$ *are, respectively, the time (in units of second) to generate one source symbol* U_ℓ *and the time to transmit one channel symbol* X_ℓ*. On the other hand, if*

$$\frac{1}{\mathrm{T_s}} H(U) \text{ bits/second} > \frac{1}{\mathrm{T_c}} \max_{P_X} I(X;Y) \text{ bits/second}, \tag{7.2}$$

then $\Pr\left[\hat{\mathbf{U}}^k \neq \mathbf{U}^k\right]$ *has a universal positive lower bound for all coding schemes of any length k; hence, the error cannot be made arbitrarily small.*

Recall from Definition 6.20 that the rate of a code is defined as

$$\mathrm{R} = \frac{\log_2 \mathrm{M}}{n} \text{ bits per transmission.} \tag{7.3}$$

From Figure 7.1 we now see that here we have[1] $\mathrm{M} = 2^k$, i.e.

$$\mathrm{R} = \frac{\log_2(2^k)}{n} = \frac{k}{n}. \tag{7.4}$$

On the other hand, it is also apparent from Figure 7.1 that, due to timing reasons, we must have

$$k\mathrm{T_s} = n\mathrm{T_c}. \tag{7.5}$$

[1] The number of codewords M is given by the number of different source sequences $\mathbf{U}^k$ of length k. For simplicity we assume here that the source is binary, i.e. $|\mathcal{U}| = 2$. In general we have $\mathrm{M} = |\mathcal{U}|^k$ and $\mathrm{R} = (k/n)\log_2|\mathcal{U}|$.

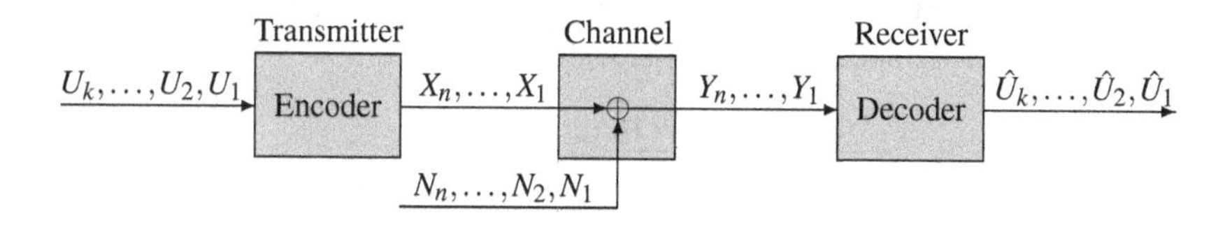

Figure 7.1 Noisy information transmission system.

Combined with (7.4) we thus see that the code rate can also be represented as

$$\mathsf{R} = \frac{\mathsf{T_c}}{\mathsf{T_s}}. \tag{7.6}$$

In Sections 7.3 and 7.4, we will further explore the two situations corresponding to whether condition (7.1) or (7.2) is valid.

7.2 The Gaussian channel

Figure 7.1 considers the situation of a binary source $U_1, U_2, \ldots, U_k$ being transmitted through a noisy channel that is characterized by

$$Y_\ell = X_\ell + N_\ell, \quad \ell = 1, 2, \ldots, n, \tag{7.7}$$

where $X_1, X_2, \ldots, X_n$ and $Y_1, Y_2, \ldots, Y_n$ are, respectively, the channel input and channel output sequences, and $N_1, N_2, \ldots, N_n$ is the noise sequence.

Assume that U_ℓ is either 0 or 1, and $\Pr[U_\ell = 0] = \Pr[U_\ell = 1] = 1/2$. Also assume that $U_1, U_2, \ldots, U_k$ are all independent.[2] Hence, its average entropy is equal to

$$\frac{1}{k}H(\mathbf{U}^k) = \frac{1}{k}\sum_{\mathbf{u}^k \in \{0,1\}^k} \Pr\left[\mathbf{U}^k = \mathbf{u}^k\right] \log_2\left(\frac{1}{\Pr\left[\mathbf{U}^k = \mathbf{u}^k\right]}\right) \text{ bits/source symbol} \tag{7.8}$$

$$= \frac{1}{k}\sum_{\mathbf{u}^k \in \{0,1\}^k} 2^{-k}\log_2\left(\frac{1}{2^{-k}}\right) \text{ bits/source symbol} \tag{7.9}$$

$$= 1 \text{ bit/source symbol}, \tag{7.10}$$

where we abbreviate $(U_1, U_2, \ldots, U_k)$ as $\mathbf{U}^k$.

In a practical communication system, there usually exists a certain constraint

[2] This assumption is well justified in practice: were U_ℓ not uniform and independent, then any good data compression scheme could make them so. For more details on this, we refer to Appendix 7.7.

E on the transmission power (for example, in units of joule per transmission). This power constraint can be mathematically modeled as

$$\frac{x_1^2 + x_2^2 + \cdots + x_n^2}{n} \leq \mathsf{E} \quad \text{for all } n. \tag{7.11}$$

When being transformed into an equivalent statistical constraint, one can replace (7.11) by

$$\mathsf{E}\left[\frac{X_1^2 + X_2^2 + \cdots + X_n^2}{n}\right] = \mathsf{E}\left[X_1^2\right] \leq \mathsf{E}, \tag{7.12}$$

where $\mathsf{E}[\cdot]$ denotes the expected value of the target random variable, and equality holds because we assume $\mathsf{E}\left[X_1^2\right] = \mathsf{E}\left[X_2^2\right] = \cdots = \mathsf{E}\left[X_n^2\right]$, i.e. the channel encoder is expected to assign, on average, an equal transmission power to each channel input. Note that the channel inputs are in general strongly dependent so as to combat interference; what we assume here is that they have, on average, equal marginal power. We further assume that the noise samples $N_1, N_2, \ldots, N_n$ are independent in statistics and that the probability density function[3] of each N_ℓ is given by

$$f_{N_\ell}(t) = \frac{1}{\sqrt{2\pi\sigma^2}} \exp\left(-\frac{t^2}{2\sigma^2}\right), \quad t \in \Re. \tag{7.13}$$

This is usually termed the *zero-mean Gaussian distribution*, and the corresponding channel (7.7) is therefore called the *Gaussian channel*.

7.3 Transmission at a rate below capacity

It can be shown that the channel capacity (i.e. the largest code rate below which arbitrarily small error probability can be obtained) of a Gaussian channel as defined in Section 7.2 is

$$\mathsf{C}(\mathsf{E}) = \max_{f_X : \mathsf{E}[X^2] \leq \mathsf{E}} I(X;Y) \tag{7.14}$$

$$= \frac{1}{2}\log_2\left(1 + \frac{\mathsf{E}}{\sigma^2}\right) \text{ bits/channel use,} \tag{7.15}$$

[3] A *probability density function* is the *density* function for *probability*. Similar to the fact that the density of a material describes its mass per unit volume, the probability density function gives the probability of occurrence per unit point. Hence, integration of the material density over a volume leads to the mass confined within it, and integration of the probability density function over a range tells us the probability that one will observe a value in this range. The Gaussian density function in (7.13) is named after the famous mathematician Carl Friedrich Gauss, who used it to analyze astronomical data. Since it is quite commonly seen in practice, it is sometimes named the *normal* distribution.

where the details can be found in [CT06, Eq. (9.16)]. Recall from the Information Transmission Theorem (Theorem 7.1) that if the source entropy (namely, 1 bit/source symbol) is less than the capacity of a communication channel (i.e. $(1/2)\log_2(1+\mathsf{E}/\sigma^2)$ bits per channel use), then reliable transmission becomes feasible. Hence, in equation form, we can present the condition for reliable transmission as follows:

$$\frac{1}{\mathsf{T}_s}\text{ bits/second} < \frac{1}{2\mathsf{T}_c}\log_2\left(1+\frac{\mathsf{E}}{\sigma^2}\right)\text{ bits/second.} \tag{7.16}$$

Note that when comparing we have to represent the average source entropy and channel capacity by the same units (here, bits/second). This is the reason why we have introduced T_s and T_c.

7.4 Transmission at a rate above capacity

Now the question is what if

$$\frac{1}{\mathsf{T}_s}\text{ bits/second} > \frac{1}{2\mathsf{T}_c}\log_2\left(1+\frac{\mathsf{E}}{\sigma^2}\right)\text{ bits/second.} \tag{7.17}$$

In such a case, we know from the Information Transmission Theorem (Theorem 7.1) that an arbitrarily small error probability cannot be achieved. However, can we identify the smallest error rate that can be possibly obtained?

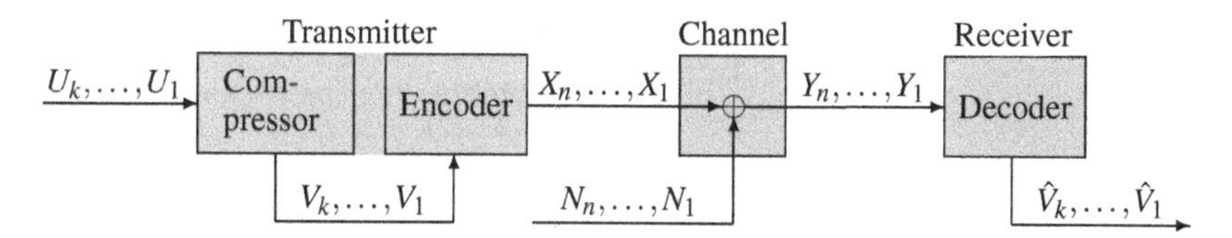

Figure 7.2 Noisy information transmission system with incorporated compressor.

A straightforward system design is to map each source sequence $u_1,\ldots,u_k$ into a compressed binary sequence $\mathbf{v}^k \triangleq g(\mathbf{u}^k)$ for transmission (see Figure 7.2), where the compressor function $g(\cdot)$ is chosen such that the resulting average entropy is less than the channel capacity, i.e.

$$\frac{1}{\mathsf{T}_s}\frac{1}{k}H(\mathbf{V}^k)\text{ bits/second} < \frac{1}{2\mathsf{T}_c}\log_2\left(1+\frac{\mathsf{E}}{\sigma^2}\right)\text{ bits/second.} \tag{7.18}$$

Then, from Theorem 7.1 we know that $V_1,V_2,\ldots,V_k$ can be transmitted through the additive Gaussian noise channel with arbitrarily small error. This is to say, there exists a transmission scheme such that the decision at the channel output $\hat{v}_1, \hat{v}_2, \hat{v}_3, \ldots$ is the same as the compressed channel input $v_1, v_2, v_3, \ldots$ with probability arbitrarily close to unity.[4] As a consequence, the error that is introduced in this straightforward system design occurs only at those instances (i.e. ℓs) where the compression is not reversible, i.e. u_ℓ cannot be recovered from $\mathbf{v}^k$.

In the following, we will determine a minimum error probability that the straightforward system in Figure 7.2 cannot beat even if we optimize over all possible compressor designs subject to the average entropy of the compressor output being smaller than the channel capacity. Before we come to this analysis, we give several examples of how a compressor works and what is its error.

Example 7.2 For example, let $g(\cdot)$ map $u_1, u_2, u_3, \ldots$ into $v_1, v_2, v_3, \ldots$ in a fashion that

$$(v_1, v_2, v_3, v_4, \ldots) = g(u_1, u_2, u_3, u_4, \ldots) = (u_1, u_1, u_3, u_3, \ldots), \tag{7.19}$$

i.e. $v_{2\ell-1} = v_{2\ell} = u_{2\ell-1}$ (see Table 7.1). Since $V_{2\ell-1} = V_{2\ell}$ for every ℓ, no new information is provided by $V_{2\ell}$ given $V_{2\ell-1}$. The average entropy of $V_1, V_2, \ldots, V_k$, with $k = 2m$ even, is then given by

$$\frac{1}{k}H(\mathbf{V}^k) = \frac{1}{2m}H(V_1, V_2, \ldots, V_{2m}) \tag{7.20}$$

$$= \frac{1}{2m}H(V_1, V_3, \ldots, V_{2m-1}) \tag{7.21}$$

$$= \frac{1}{2m}H(U_1, U_3, U_5, \ldots, U_{2m-1}) \tag{7.22}$$

$$= \frac{1}{2} \text{ bits/source symbol}, \tag{7.23}$$

i.e.

$$\frac{1}{\mathrm{T_s}}\frac{1}{k}H(\mathbf{V}^k) = \frac{1}{2\mathrm{T_s}} \text{ bits/second.} \tag{7.24}$$

[4] What Shannon targeted in his theorem is the *block error rate*, not the *bit error rate*. Hence, his theorem actually concludes that $\Pr\left[(V_1, V_2, \ldots, V_k) = (\hat{V}_1, \hat{V}_2, \ldots, \hat{V}_k)\right] \simeq 1$ when k is sufficiently large since

$$\frac{1}{\mathrm{T_s}}\frac{1}{k}H(\mathbf{V}^k) \text{ bits/second}$$

is less than

$$\frac{1}{2\mathrm{T_c}}\log_2\left(1 + \frac{\mathrm{E}}{\sigma^2}\right) \text{ bits/second.}$$

This is a very strong statement because, for example, $v_\ell = \hat{v}_\ell$ for $1 \le \ell \le k-1$ and $v_k \ne \hat{v}_k$ will be counted as one block error even though there is only one bit error among these k bits. Note that $\Pr\left[(V_1, V_2, \ldots, V_k) = (\hat{V}_1, \hat{V}_2, \ldots, \hat{V}_k)\right] \simeq 1$ surely implies that $\Pr\left[V_\ell = \hat{V}_\ell\right] \simeq 1$ for most ℓ, but not vice versa.

Table 7.1 *The mapping $g(\cdot)$ from $\mathbf{u}^k$ to $\mathbf{v}^k$ defined in (7.19) in Example 7.2*

$u_1, u_2, u_3, u_4, \ldots$	$v_1, v_2, v_3, v_4, \ldots$
0000...	0000...
0001...	0000...
0010...	0011...
0011...	0011...
0100...	0000...
0101...	0000...
0110...	0011...
0111...	0011...
1000...	1100...
1001...	1100...
1010...	1111...
1011...	1111...
1000...	1100...
1101...	1100...
1110...	1111...
1111...	1111...

Under the premise that

$$\frac{1}{2\mathsf{T}_\mathrm{s}} \text{ bits/second} < \frac{1}{2\mathsf{T}_\mathrm{c}} \log_2\left(1 + \frac{\mathsf{E}}{\sigma^2}\right) \text{ bits/second}, \tag{7.25}$$

the average "bit" error rate of the system is given by

$$\begin{aligned}
&\frac{1}{k}\Big(\Pr[U_1 \neq V_1] + \Pr[U_2 \neq V_2] + \Pr[U_3 \neq V_3] + \Pr[U_4 \neq V_4] \\
&\quad + \cdots + \Pr[U_k \neq V_k]\Big) \\
&= \frac{1}{2m}\Big(\Pr[U_1 \neq V_1] + \Pr[U_2 \neq V_2] + \Pr[U_3 \neq V_3] + \Pr[U_4 \neq V_4] \\
&\qquad + \cdots + \Pr[U_{2m} \neq V_{2m}]\Big) &(7.26)\\
&= \frac{1}{2m}\Big(0 + \Pr[U_2 \neq V_2] + 0 + \Pr[U_4 \neq V_4] + \cdots + \Pr[U_{2m} \neq V_{2m}]\Big) &(7.27)\\
&= \frac{1}{2m}\Big(\Pr[U_2 \neq U_1] + \Pr[U_4 \neq U_3] + \cdots + \Pr[U_{2m} \neq U_{2m-1}]\Big) &(7.28)\\
&= \frac{1}{2}\Pr[U_2 \neq U_1] &(7.29)\\
&= \frac{1}{2}\Big(\Pr[(U_1, U_2) = (01)] + \Pr[(U_1, U_2) = (10)]\Big) &(7.30)
\end{aligned}$$

$$= \frac{1}{2}\left(\frac{1}{4}+\frac{1}{4}\right) = \frac{1}{4}. \tag{7.31}$$

In fact, the error rate of $1/4$ can be obtained directly without computation by observing that all the odd-indexed bits $U_1, U_3, U_5, \ldots$ can be correctly recovered from $V_1, V_3, V_5, \ldots$; however, the sequence $\mathbf{V}^k$ provides no information for all the even-indexed bits $U_2, U_2, U_6, \ldots$ Thus, we can infer that a zero error rate for odd-indexed bits and a $1/2$ error rate based on a pure guess for even-indexed bits combine to a $1/4$ error rate. ◊

Example 7.2 provides a compressor with average bit error rate (BER) of $1/4$ between input $U_1, U_2, \ldots, U_k$ and output $V_1, V_2, \ldots, V_k$ subject to its average output entropy $1/2\mathrm{T_s}$ bits/second being smaller than the channel capacity $\mathsf{C}(\mathsf{E}) = (1/2\mathrm{T_c})\log_2(1+\mathsf{E}/\sigma^2)$. However, dropping all even-indexed bits may not be a good compressor design because it is possible that half of the bits in $V_1, V_2, \ldots, V_k$ are different from $U_1, U_2, \ldots, U_k$; i.e., in the worst case, the difference between compressor input $U_1, U_2, \ldots, U_k$ and compressor output $V_1, V_2, \ldots, V_k$ will result in a large distortion of $k/2$ bits.

In Section 3.3.3 we saw that the $(7,4)$ Hamming code is a perfect packing of radius-one spheres in the 7-dimensional binary space. Using this property, we can provide in Example 7.3 an alternative compressor design such that the input and output are different in at most 1 bit with a (smaller) average BER of $1/8$ and a (slightly larger) average output entropy of $4/7\mathrm{T_s}$ bits/second.

Example 7.3 A compressor g is defined based on the $(7,4)$ Hamming codewords listed in Table 3.2 as follows: $g(\mathbf{u}^7) = \mathbf{v}^7$ if $\mathbf{v}^7$ is a $(7,4)$ Hamming codeword and $\mathbf{u}^7$ is at Hamming distance at most one from $\mathbf{v}^7$. The perfect packing of the 16 nonoverlapping radius-one spheres centered at the codewords for the $(7,4)$ Hamming code guarantees the existence and uniqueness of such a $\mathbf{v}^7$; hence, the compressor function mapping is well defined.

The probability of each $(7,4)$ Hamming codeword appearing at the output is $8 \cdot 2^{-7} = 2^{-4}$ (since there are eight $\mathbf{u}^7$ mapped to the same $\mathbf{v}^7$). Hence,

$$\frac{1}{7}H(\mathbf{V}^7) = \frac{1}{7}\sum_{\mathbf{v}^7 \in \mathscr{C}_\mathrm{H}} 2^{-4}\log_2\left(\frac{1}{2^{-4}}\right) \tag{7.32}$$

$$= \frac{4}{7} \text{ bits/source symbol,} \tag{7.33}$$

where $\mathscr{C}_\mathrm{H}$ denotes the set of the 16 codewords of the $(7,4)$ Hamming code. Hence,

$$\frac{1}{7\mathrm{T_s}}H(\mathbf{V}^7) = \frac{4}{7\mathrm{T_s}} \text{ bits/second.} \tag{7.34}$$

Next, we note that $\Pr[U_1 \neq V_1] = 1/8$ because only one of the eight $\mathbf{u}^7$ that are mapped to the same $\mathbf{v}^7$ results in a different first bit. Similarly, we can obtain

$$\Pr[U_2 \neq V_2] = \Pr[U_3 \neq V_3] = \cdots = \Pr[U_7 \neq V_7] = \frac{1}{8}. \tag{7.35}$$

Hence, the average BER is given by

$$\text{BER} = \frac{1}{7}\big(\Pr[U_1 \neq V_1] + \Pr[U_2 \neq V_2] + \cdots \Pr[U_7 \neq V_7]\big) = \frac{1}{8}. \tag{7.36}$$

This example again shows that data compression can be regarded as the opposite operation of error correction coding, where the former removes the redundancy (or even some information such as in this example) while the latter adds controlled redundancy to combat the channel noise effect. ◇

Exercise 7.4 *Design a compressor mapping by reversing the roles of encoder and decoder of the three-times repetition code. Prove that the average BER is 1/4 and the average output entropy equals 1/3 bits per source symbol.* ◇

Readers may infer that one needs to know the best compressor design, which minimizes the BER subject to the average output entropy less than $\mathsf{C}(\mathsf{E})$, in order to know what will be the minimum BER attainable for a given channel (or more specifically for a given $\mathsf{C}(\mathsf{E})$). Ingeniously, Shannon identifies this minimum BER without specifying how it can be achieved. We will next describe his idea of a converse proof that shows that the minimum BER cannot be smaller than some quantity, but that does not specify $g(\cdot)$.

We can conceptually treat the compressor system as a channel with input $U_1, U_2, \ldots, U_k$ and output $V_1, V_2, \ldots, V_k$. Then, by

$$H(\mathbf{V}^k|\mathbf{U}^k) = H(g(\mathbf{U}^k)|\mathbf{U}^k) = 0, \tag{7.37}$$

we derive from (6.85) that

$$I(\mathbf{U}^k;\mathbf{V}^k) = H(\mathbf{U}^k) - H(\mathbf{U}^k|\mathbf{V}^k) = H(\mathbf{V}^k) - H(\mathbf{V}^k|\mathbf{U}^k) = H(\mathbf{V}^k). \tag{7.38}$$

This implies that the average entropy of the compressor output is equal to

$$\frac{1}{k}H(\mathbf{V}^k) = \frac{1}{k}H(\mathbf{U}^k) - \frac{1}{k}H(\mathbf{U}^k|\mathbf{V}^k) = 1 - \frac{1}{k}H(\mathbf{U}^k|\mathbf{V}^k) \text{ bits.} \tag{7.39}$$

By the chain rule for entropy,[5]

$$\frac{1}{k}H(\mathbf{V}^k) = 1 - \frac{1}{k}H(\mathbf{U}^k|\mathbf{V}^k) \tag{7.40}$$

[5] The chain rule for entropy is

$$H(\mathbf{U}^k) = H(U_1) + H(U_2|U_1) + H(U_3|U_1,U_2) + \cdots + H(U_k|U_1,\ldots,U_{k-1}).$$

This is a generalized form of Proposition 6.4 and can be proven similarly.

$$= 1 - \frac{1}{k}\big(H(U_1|\mathbf{V}^k) + H(U_2|U_1,\mathbf{V}^k) + H(U_3|U_1,U_2,\mathbf{V}^k) + \cdots + H(U_k|U_1,\ldots,U_{k-1},\mathbf{V}^k)\big) \quad (7.41)$$

$$\geq 1 - \frac{1}{k}\big(H(U_1|V_1) + H(U_2|V_2) + H(U_3|V_3) + \cdots + H(U_k|V_k)\big), \quad (7.42)$$

where (7.42) holds because additional information always helps to decrease entropy; i.e., $H(U_\ell|U_1,\ldots,U_{\ell-1},V_1,\ldots,V_k) \leq H(U_\ell|V_\ell)$ since the former has additional information $(U_1,\ldots,U_{\ell-1},V_1,\ldots,V_{\ell-1},V_{\ell+1},\ldots,V_k)$ (see Corollary 6.10).

We proceed with the derivation by pointing out that

$$H(U_\ell|V_\ell) = \Pr[V_\ell = 0]\, H(U_\ell|V_\ell = 0) + \Pr[V_\ell = 1]\, H(U_\ell|V_\ell = 1) \quad (7.43)$$

$$= \Pr[V_\ell = 0]\, H_\mathrm{b}\big(\Pr[U_\ell = 1 \,|\, V_\ell = 0]\big) + \Pr[V_\ell = 1]\, H_\mathrm{b}\big(\Pr[U_\ell = 0 \,|\, V_\ell = 1]\big) \quad (7.44)$$

$$\leq H_\mathrm{b}\Big(\Pr[V_\ell = 0]\Pr[U_\ell = 1 \,|\, V_\ell = 0] + \Pr[V_\ell = 1]\Pr[U_\ell = 0 \,|\, V_\ell = 1]\Big) \quad (7.45)$$

$$= H_\mathrm{b}(\mathrm{BER}_\ell), \quad (7.46)$$

where $\mathrm{BER}_\ell \triangleq \Pr[U_\ell \neq V_\ell]$;

$$H_\mathrm{b}(p) \triangleq p\log_2\frac{1}{p} + (1-p)\log_2\frac{1}{1-p}, \qquad \text{for } 0 \leq p \leq 1, \quad (7.47)$$

is the so-called *binary entropy function* (see Section 5.2.2); and (7.45) follows from the concavity[6] of the function $H_\mathrm{b}(\cdot)$. We then obtain the final lower bound of the output average entropy:

$$\frac{1}{k}H(\mathbf{V}^k) \geq 1 - \frac{1}{k}\big(H_\mathrm{b}(\mathrm{BER}_1) + H_\mathrm{b}(\mathrm{BER}_2) + \cdots + H_\mathrm{b}(\mathrm{BER}_k)\big) \quad (7.48)$$

$$= 1 - \frac{1}{k}\sum_{\ell=1}^{k} H_\mathrm{b}(\mathrm{BER}_\ell) \quad (7.49)$$

$$\geq 1 - H_\mathrm{b}\left(\frac{1}{k}\sum_{\ell=1}^{k} \mathrm{BER}_\ell\right) \quad (7.50)$$

$$= 1 - H_\mathrm{b}(\mathrm{BER}). \quad (7.51)$$

Here, (7.50) follows again from concavity.

In conclusion, the Information Transmission Theorem (Theorem 7.1) identifies the achievable *bit error rate (BER)* for the target additive Gaussian noise

[6] For a definition of concavity see Appendix 7.8.

channel as follows:

$$\frac{1}{\mathsf{T_s}}(1 - H_\mathrm{b}(\mathrm{BER})) \le \frac{1}{\mathsf{T_s}}\frac{1}{k}H(\mathbf{V}^k) < \frac{1}{2\mathsf{T_c}}\log_2\left(1 + \frac{\mathsf{E}}{\sigma^2}\right), \tag{7.52}$$

where the first inequality follows from (7.51) and the second follows from our assumption (7.18). In usual communication terminologies, people denote $\mathsf{R} = \mathsf{T_c}/\mathsf{T_s} = k/n$ (information bit carried per channel use) as the *channel code rate*; $\mathsf{N_0} \triangleq 2\sigma^2$ (joule) as the *noise energy level*; $\mathsf{E_b} \triangleq \mathsf{ET_s}/\mathsf{T_c}$ (joule) as the *equivalent transmitted energy per information bit*; and $\gamma_\mathrm{b} \triangleq \mathsf{E_b}/\mathsf{N_0}$ as the *signal-to-noise power ratio per information bit*. This transforms the above inequality to

$$H_\mathrm{b}(\mathrm{BER}) > 1 - \frac{1}{2\mathsf{R}}\log_2(1 + 2\mathsf{R}\gamma_\mathrm{b}). \tag{7.53}$$

Equation (7.53) clearly indicates that the BER cannot be made smaller than

$$H_\mathrm{b}^{-1}\left(1 - \frac{1}{2\mathsf{R}}\log_2(1 + 2\mathsf{R}\gamma_\mathrm{b})\right), \tag{7.54}$$

where $H_\mathrm{b}^{-1}(\cdot)$ is the inverse function of the binary entropy function $H_\mathrm{b}(\xi)$ (see Section 5.2.2) for $\xi \in [0, 1/2]$. Shannon also proved the (asymptotic) achievability of this lower bound. Hence, (7.53) provides the exact margin on what we can do and what we cannot do when the amount of information to be transmitted is above the capacity.

We plot the curves corresponding to $\mathsf{R} = 1/2$ and $\mathsf{R} = 1/3$ in Figure 7.3. The figure indicates that there exists a rate-$1/2$ system design that can achieve $\mathrm{BER} = 10^{-5}$ at $\gamma_{\mathrm{b,dB}} \triangleq 10\log_{10}(\mathsf{E_b}/\mathsf{N_0})$ close to 0 dB, i.e. for $\mathsf{E_b} \simeq \mathsf{N_0}$. On the other hand, no system with a rate $\mathsf{R} = 1/2$ can yield a BER less than 10^{-5} if the signal energy per information bit $\mathsf{E_b}$ is less than the noise energy level $\mathsf{N_0}$. Information theorists therefore call this threshold the *Shannon limit*.

For decades (ever since Shannon ingeniously drew such a sacred line in 1948 simply by analysis), researchers have tried to find a good design that can achieve the Shannon limit. Over the years, the gap between the real transmission scheme and this theoretical limit has been gradually closed. For example, a concatenated code [For66] proposed by David Forney can reach $\mathrm{BER} = 10^{-5}$ at about $\gamma_{\mathrm{b,dB}} \simeq 2$ dB. However, no schemes could push their performance curves within 1 dB of the Shannon limit until the invention of turbo codes in 1993 [BGT93]. Motivated by the turbo coding idea, the low-density parity-check (LDPC) codes were subsequently rediscovered[7] in 1998; these could

[7] We use "rediscover" here because the LDPC code was originally proposed by Robert G. Gallager in 1962 [Gal62]. However, due to its high complexity, computers at that time could not perform any simulations on the code; hence, nobody realized the potential of LDPC codes. It

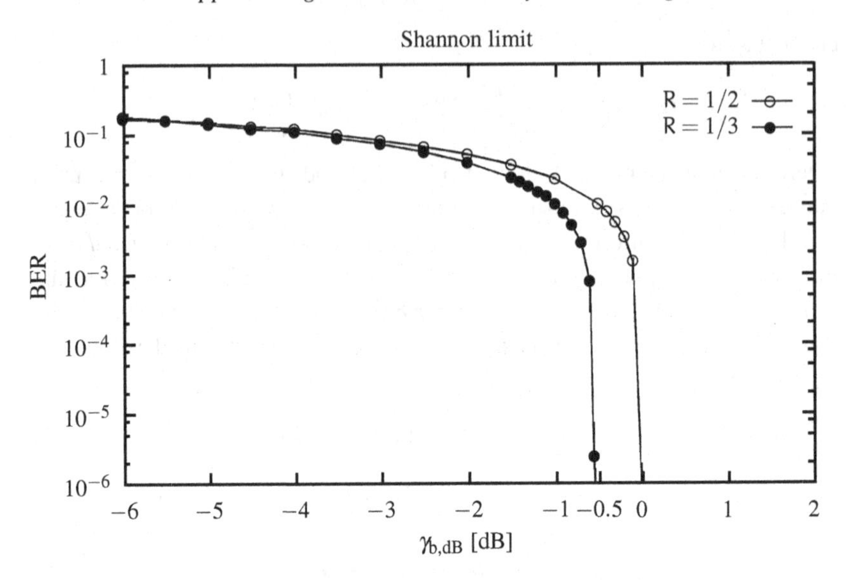

Figure 7.3 The Shannon limits for rates $1/2$ and $1/3$ codes on continuous-input AWGN channels. Decibel (abbreviated as dB) is a logarithmic scaling of a given quantity; i.e., we first take the base-10 logarithm and then multiply by 10. So, e.g., $\gamma_{b,dB} \triangleq 10\log_{10}(\gamma_b) = 10\log_{10}(E_b/N_0)$.

reduce the performance gap (between the LDPC codes and the Shannon limit) within, e.g., 0.1 dB. This counts 50 years of efforts (from 1948 to 1998) until we finally caught up with the pioneering prediction of Shannon in the classical additive Gaussian noise channel.

With excitement, we should realize that this is just the beginning of closing the gap, not the end of it. Nowadays, the channels we face in reality are much more complicated than the simple additive Gaussian noise channel. Multipath and fading effects, as well as channel nonlinearities, make the Shannon-limit approaching mission in these channels much more difficult. New ideas other than turbo and LDPC coding will perhaps be required in the future. So we are waiting for some new exciting results, similar to the discovery of turbo codes in 1993.

was Matthew Davey and David MacKay who rediscovered and examined the code, and confirmed its superb performance in 1998 [DM98].

7.5 Turbo coding: an introduction

Of all error-correcting codes to date, the turbo code was the first that could approach the Shannon limit within 1 dB, at BER $= 10^{-5}$, over the additive Gaussian noise channel. It is named the *turbo code* because the decoder functions iteratively like a turbo machine, where two turbo engines take turns to refine the previous output of the other until a certain number of iterations is reached.

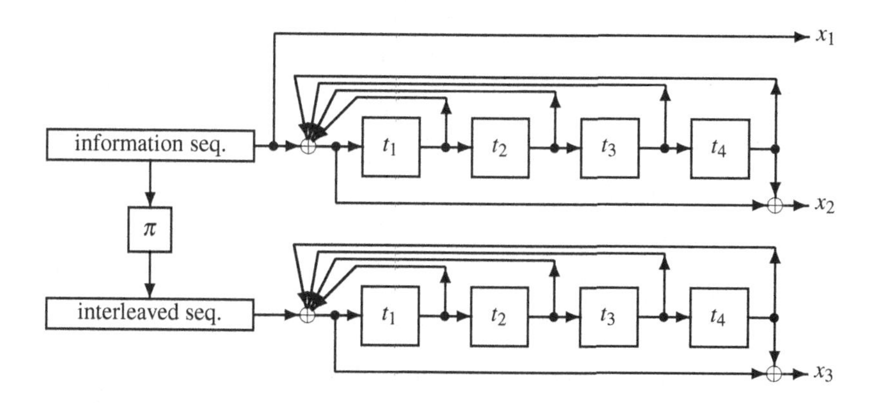

Figure 7.4 Exemplified turbo encoder from [BGT93]. An example of how a length-5 input sequence passes through this encoder is depicted in Figure 7.5. The complete list of all length-5 input sequences with their corresponding codewords is given in Table 7.2.

As an example, an information sequence $(s_1, s_2, s_3, s_4, s_5) = (10100)$ is fed into the turbo encoder shown in Figure 7.4, where s_1 is inserted first. In this figure, the squares marked with t_1, t_2, t_3, and t_4 are clocked memory elements, usually named *flip-flops* or simply *registers*, which store the coming input binary data and, at the same time, output its current content according to a clocked timer. The square marked with π is the interleaver that permutes the input sequence into its interleaved counterpart. The notation "$\oplus$" denotes the modulo-2 addition.

This figure then indicates that the output sequence from node x_1 will be the original information sequence $(s_1, s_2, s_3, s_4, s_5) = (10100)$. Since the contents of all four registers are initially zero, and since

$$t_1(\ell+1) = s_\ell \oplus t_1(\ell) \oplus t_2(\ell) \oplus t_3(\ell) \oplus t_4(\ell), \tag{7.55}$$

input sequence	register contents				output sequence
$s_5s_4s_3s_2s_1 \rightarrow$	t_1	t_2	t_3	t_4	$\rightarrow x_2(5)x_2(4)x_2(3)x_2(2)x_2(1)$
00101	0	0	0	0	
0010	1	0	0	0	1
001	1	1	0	0	11
00	1	1	1	0	111
0	1	1	1	1	1111
	0	1	1	1	11111

Figure 7.5 The snap show of the input and output sequences of the turbo encoder from Figure 7.4 at node x_2. Note that we have mirrored the sequences to match the direction of the register placement in Figure 7.4.

$$t_2(\ell+1) = t_1(\ell), \tag{7.56}$$

$$t_3(\ell+1) = t_2(\ell), \tag{7.57}$$

$$t_4(\ell+1) = t_3(\ell), \tag{7.58}$$

$$x_2(\ell) = t_1(\ell+1) \oplus t_4(\ell) = s_\ell \oplus t_1(\ell) \oplus t_2(\ell) \oplus t_3(\ell), \tag{7.59}$$

where ℓ represents the clocked time instance, we obtain the output sequence from node x_2 as 11111 (see Figure 7.5). Note that in order to start indexing all sequences from 1, we re-adjust the index at the output such that $x_2(\ell)$ is actually outputted at clocked time instance $\ell + 1$.

An important feature of the turbo code design is the incorporation of an interleaver π that permutes the input sequence. For example,

$$\pi(s_1s_2s_3s_4s_5) = s_4s_2s_1s_5s_3. \tag{7.60}$$

In concept, the purpose of adding an interleaver is to introduce distant dependencies into the codewords. Notably, a strong dependency among the code bits can greatly enhance their capability against local noisy disturbance. A good example is a code of two codewords, i.e.

$$000000000000000 \quad \text{and} \quad 111111111111111,$$

where the code bits are all the same and hence are strongly dependent. Since the receiver knows all code bits should be the same, the local noisy disturbances that alter, for example, code bits 3, 7, and 10, yielding

$$001000100100000,$$

Table 7.2 *The information sequences of length 5 and their respective turbo codewords of length 15 for the turbo code in Figure 7.4 with interleaver* $\pi(s_1s_2s_3s_4s_5) = s_4s_2s_1s_5s_3$

Information sequences	Codewords		
$\mathbf{s}$	$\mathbf{x}_1$	$\mathbf{x}_2$	$\mathbf{x}_3$
$s_1s_2s_3s_4s_5$	$\mathbf{x}_1 = \mathbf{s}$	$x_2(1)x_2(2)x_2(3)x_2(4)x_2(5)$	$x_3(1)x_3(2)x_3(3)x_3(4)x_3(5)$
00000	—	00000	00000
00001	—	00001	00011
00010	—	00011	11001
00011	—	00010	11010
00100	—	00110	00001
00101	—	00111	00010
00110	—	00101	11000
00111	—	00100	11011
01000	—	01100	01100
01001	—	01101	01111
01010	—	01111	10101
01011	—	01110	10110
01100	—	01010	01101
01101	—	01011	01110
01110	—	01001	10100
01111	—	01000	10111
10000	—	11001	00110
10001	—	11000	00101
10010	—	11010	11111
10011	—	11011	11100
10100	—	11111	00111
10101	—	11110	00100
10110	—	11100	11110
10111	—	11101	11101
11000	—	10101	01010
11001	—	10100	01001
11010	—	10110	10011
11011	—	10111	10000
11100	—	10011	01011
11101	—	10010	01000
11110	—	10000	10010
11111	—	10001	10001

can be easily recovered back to the transmitted codeword

$$000000000000000.$$

By interleaving the information bits, s_1 may now affect distant code bits such as $x_3(\ell)$, where ℓ can now be much larger than the number of registers, 4. This is contrary to the conventional coding scheme for which the code bit is only a function of several recent information bits. For example, without interleaving, s_1 can only affect $x_2(4)$ but not any $x_2(\ell)$ with $\ell > 4$ according to (7.55)–(7.59). We can of course purposely design a code such that each code bit is a function of more distant information bits, but the main problem here is that the strong dependency of code bits on distant information bits will make the decoder infeasibly complex.

This leads to another merit of using the interleaver: it helps structure a feasible decoding scheme, i.e. turbo decoding. In short, the sub-decoder based on $\mathbf{x}_1$ and $\mathbf{x}_2$ will deal with a code that only has local dependencies as each code bit only depends on the previous four information bits. The second sub-decoder based on interleaved $\mathbf{x}_1$ and $\mathbf{x}_3$ similarly handles a code with only local dependencies. By this design, the task of decoding the code with distant dependencies can be accomplished by the cooperation of two feasible sub-decoders.

To be specific, the practice behind turbo decoding is to first decode the information sequence based on the noisy receptions due to the transmission of sequences $\mathbf{x}_1$ and $\mathbf{x}_2$ (in terms of the above example, 10100 and 11111). Since the code bits generated at node x_2 depend on previous information bits only through the contents of four registers, the sub-decoding procedure is feasible. The decoding output sequence, however, is not the final estimate about the information sequence 10100, but a sequence of real numbers that represent the probability for each bit to be, e.g. 1, calculated based on the noisy receptions due to the transmission of $\mathbf{x}_1$ and $\mathbf{x}_2$. Continuing with the example of the simple 5-bit input sequence, the decoding output sequence might be $(0.8, 0.2, 0.7, 0.1, 0.1)$. Based on these numbers, we know that with 80% probability the first bit is 1. Also, we assert with only 20% confidence that the second information bit is 1. Please note that if there is no noise during the transmission, the five real numbers in the decoding output sequence should be $(1.0, 0.0, 1.0, 0.0, 0.0)$. It is due to the noise that the receiver can only approximate the sequence that the transmitter sends. In terminology, we call these real numbers the *soft decoding outputs* in contrast to the conventional zero–one *hard decoding outputs*.

After obtaining the real-valued decoding output based on the noisy receptions due to the transmission of $\mathbf{x}_1$ and $\mathbf{x}_2$, one can proceed to refine these numbers by performing a similar decoding procedure based on the noisy receptions due to the transmission of $\mathbf{x}_1$ and $\mathbf{x}_3$ as well as the interleaved soft decoding output from the previous step, e.g. $(0.1, 0.2, 0.8, 0.1, 0.7)$ subject to the interleaver $\pi(s_1 s_2 s_3 s_4 s_5) = s_4 s_2 s_1 s_5 s_3$. With the additional knowledge from the noisy receptions due to the transmission of $\mathbf{x}_3$, these numbers may be refined to, e.g., $(0.05, 0.1, 0.9, 0.05, 0.8)$; hence, the receiver is more certain (here, 90% sure) that s_1 should be 1.

By performing the decoding procedure based on the noisy receptions due to $\mathbf{x}_1$ and $\mathbf{x}_2$ as well as the de-interleaved soft decoding output (e.g. $(0.9, 0.1, 0.8, 0.05, 0.05)$), these numbers are refined again. Then, in terms of these re-refined numbers, the decoding procedure based on the noisy receptions due to $\mathbf{x}_1$ and $\mathbf{x}_3$ is re-performed.

Because the repetitive decoding procedures are similar to running two turbo pistons alternatively, it is named *turbo coding*. Simulations show that after 18 iterations we can make the final hard decisions (i.e. 0 and 1 on each bit) based on the repeatedly refined soft decoding outputs yielding a bit error rate almost achieving the Shannon limit.

Ever since the publication of turbo coding, iterative decoding schemes have become a new research trend, and codes similar to the rediscovered LDPC code have subsequently been proposed. In this way the Shannon limit finally has become achievable after 50 years of research efforts!

7.6 Further reading

In this chapter we introduced the Information Transmission Theorem (Theorem 7.1) as a summary of what we have learned in this book. In order to appreciate the beauty of the theorem, we then examined it under a specific case when the coded information is corrupted by additive Gaussian noise. Two scenarios followed in a straightforward manner: transmission at a rate below capacity and transmission at a rate above capacity. The former directed us to reliable transmission where decoding errors can be made arbitrarily small, while the latter gave the (in principle) required $\mathsf{E_b}/\mathsf{N_0}$ to achieve an acceptable BER. Information theorists have baptized this minimum $\mathsf{E_b}/\mathsf{N_0}$ the *Shannon limit*. Since this is the center of information theory, most advanced textbooks in this area cover the subject extensively. For readers who are interested in learning more about the theorem, [CT06] could be a good place to start. The long-standing bible-like textbook [Gal68] by Robert G. Gallager, who was also the inventor of the Shannon-limit-achieving low-density parity-check (LDPC) codes, can also serve as a good reference. Some advanced topics in information theory, such as the channel reliability function, can be found in [Bla88].

As previously mentioned, the turbo code was the first empirically confirmed near-Shannon-limit error-correcting code. It is for this reason that the turbo code was introduced briefly at the end of this chapter. The two books [HW99] and [VY00] may be useful for those who are specifically interested in the practice and principle of turbo codes. Due to their significance, Shu Lin and Daniel Costello also devote one chapter for each of turbo coding and LDPC coding in their book in the new 2004 edition [LC04]. For general readers, the material in these two chapters should suffice for a comprehensive understanding of the related subjects.

7.7 Appendix: Why we assume uniform and independent data at the encoder

It is very common to assume that the input S_ℓ of a channel encoder comprise independent binary data bits of uniform distribution

$$\Pr[S_\ell = 0] = \Pr[S_\ell = 1] = \frac{1}{2}. \tag{7.61}$$

The reason for this lies in the observation that, under the assumption of independence with uniform marginal distribution, no data compression can be performed further on the sequence $S_1, S_2, \ldots, S_k$ since every binary combination of length k has the same probability 2^{-k} (or specifically, $(1/k)H(\mathbf{S}^k) = 1$ bit per source symbol). Note that any source that is compressed by an *optimal* data compressor should in principle produce a source output of this kind, and we can regard this assumption as that $S_1, S_2, \ldots, S_k$ are the output from an optimal data compressor.

For a better understanding of this notion, consider the following example. Assume that we wish to compress the sequence $U_1, U_2, U_3, \ldots$ to the binary sequence $S_1, S_2, S_3, \ldots$, and that each source output U_ℓ is independent of all others and has the probability distribution

$$\Pr[U_\ell = a] = \frac{1}{2}, \quad \Pr[U_\ell = b] = \Pr[U_\ell = c] = \frac{1}{4}. \tag{7.62}$$

We then use the single-letter (i.e. $\nu = 1$) Huffman code that maps as follows:

$$a \mapsto 0, \qquad b \mapsto 10, \qquad c \mapsto 11. \tag{7.63}$$

Note that (7.63) is also a Fano code (see Definition 5.17); in fact, when the source probabilities are reciprocals of integer powers of 2, both the Huffman code and the Fano code are optimal compressors with average codeword length

L_{av} equal to the source entropy, $H(U)$ bits. This then results in

$$\Pr\left[S_{\ell+1}=0 \,\middle|\, \mathbf{S}^{\ell}=\mathbf{s}^{\ell}\right] = \begin{cases} \Pr\left[U_{\ell'+1}=a \,\middle|\, \mathbf{U}^{\ell'}=\mathbf{u}^{\ell'}\right] & \text{if } \text{code}(\mathbf{u}^{\ell'})=\mathbf{s}^{\ell}, \\ \Pr\left[U_{\ell'+1}=b \,\middle|\, \mathbf{U}^{\ell'}=\mathbf{u}^{\ell'} \text{ and } U_{\ell'+1}\neq a\right] & \text{if } \text{code}(\mathbf{u}^{\ell'})=\mathbf{s}^{\ell-1} \\ & \text{and } s_{\ell}=1. \end{cases} \tag{7.64}$$

Here ℓ' denotes the symbol-timing at the input of the Huffman encoder (or, equivalently, the Fano encoder), and ℓ is the corresponding timing of the output of the Huffman encoder (equivalently, the Fano encoder).

We can now conclude by the independence of the sequence $U_1, U_2, U_3, \ldots$ that

$$\Pr\left[S_{\ell+1}=0 \,\middle|\, \mathbf{S}^{\ell}=\mathbf{s}^{\ell}\right] = \begin{cases} \Pr[U_{\ell'+1}=a] & \text{if } \text{code}(\mathbf{u}^{\ell'})=\mathbf{s}^{\ell}, \\ \Pr[U_{\ell'+1}=b \,|\, U_{\ell'+1}\neq a] & \text{if } \text{code}(\mathbf{u}^{\ell'})=\mathbf{s}^{\ell-1} \text{ and } s_{\ell}=1 \end{cases} \tag{7.65}$$

$$= \frac{1}{2}. \tag{7.66}$$

Since the resultant quantity $1/2$ is irrespective of the $\mathbf{s}^{\ell}$ given, we must have

$$\Pr\left[S_{\ell+1}=0 \,\middle|\, \mathbf{S}^{\ell}=\mathbf{s}^{\ell}\right] = \Pr[S_{\ell+1}=0] = \frac{1}{2}. \tag{7.67}$$

Hence, $S_{\ell+1}$ is independent of $\mathbf{S}^{\ell}$ and is uniform in its statistics. Since this is true for every positive integer ℓ, $S_1, S_2, S_3, \ldots$ is an independent sequence with uniform marginal distribution.

Sometimes, the output from an optimal compressor can only approach *asymptotic* independence with *asymptotic* uniform marginal distribution. This occurs when the probabilities of U are not reciprocals of powers of 2, i.e. different from what we have assumed in the previous derivation. For example, assume

$$\Pr[U=a] = \frac{2}{3} \quad \text{and} \quad \Pr[U=b] = \Pr[U=c] = \frac{1}{6}. \tag{7.68}$$

Then $S_1, S_2, S_3, \ldots$ can only be made asymptotically independent with asymptotic uniform marginal distribution in the sense that a multiple-letter code (i.e. a code that encodes several input letters at once; see Figure 5.12 in Section 5.5) needs to be used with the number of letters per compression growing to *infinity*. For example, the double-letter Huffman code in Table 7.3 gives

$$\Pr[S_1=0] = \Pr\left[\mathbf{U}^2=aa\right] = \left(\frac{2}{3}\right)^2 = \frac{4}{9} \tag{7.69}$$

Table 7.3 *Double-letter and triple-letter Huffman codes for source statistics* $\Pr[U=a]=2/3$ *and* $\Pr[U=b]=\Pr[U=c]=1/6$

Letters	Code	Letters	Code	Letters	Code	Letters	Code
aa	0	*aaa*	00	*baa*	0111	*caa*	1110
ab	100	*aab*	1100	*bab*	10100	*cab*	10110
ac	110	*aac*	0100	*bac*	10101	*cac*	10111
bb	11100	*aba*	0101	*bba*	110100	*cba*	110110
ba	1010	*abb*	10000	*bbb*	1111000	*cbb*	1111100
bc	11101	*abc*	10001	*bbc*	1111001	*cbc*	1111101
ca	1011	*aca*	0110	*bca*	110101	*cca*	110111
cb	11110	*acb*	10010	*bcb*	1111010	*ccb*	1111110
cc	11111	*acc*	10011	*bcc*	1111011	*ccc*	1111111

and

$$\Pr[S_2=0]=\Pr\left[\left(\mathbf{U}^2=ab \text{ or } ba \text{ or } ca\right) \text{ or } \mathbf{U}^4=aaaa\right] \tag{7.70}$$

$$=\frac{2}{3}\cdot\frac{1}{6}+\frac{1}{6}\cdot\frac{2}{3}+\frac{1}{6}\cdot\frac{2}{3}+\left(\frac{2}{3}\right)^4 \tag{7.71}$$

$$=\frac{43}{81}. \tag{7.72}$$

These two numbers are closer to $1/2$ than those from the single-letter Huffman code that maps a,b,c to $0,10,11$, respectively, which gives

$$\Pr[S_1=0]=\Pr[U_1\neq a]=\frac{1}{3} \tag{7.73}$$

and

$$\Pr[S_2=0]=\Pr\left[(U_1=b) \text{ or } (\mathbf{U}^2=aa)\right]=\frac{33}{54}. \tag{7.74}$$

Note that the approximation from the triple-letter Huffman code may be *transiently* less "accurate" to the uniform distribution than the double-letter Huffman code (in the current example we have $\Pr[S_1=0]=16/27$, which is less close to $1/2$ than $\Pr[S_1=0]=4/9$ from (7.69)). This complements what has been pointed out in Section 5.4.2, namely that it is rather difficult to analyze (and also rather operationally intensive to examine numerically[8]) the average

[8] As an example, when v (the number of source letters per compression) is only moderately large, such as 20, you can try to construct a Huffman code with source alphabet of size $|\{a,b,c\}|^{20}=3^{20}$ using Huffman's Algorithm from Section 4.6. Check how many iterations are required to root a tree with 3^{20} leaves.

Table 7.4 *Double-letter and triple-letter Fano codes for source statistics* $\Pr[U = a] = 2/3$ *and* $\Pr[U = b] = \Pr[U = c] = 1/6$

Letters	Code
aa	00
ab	01
ac	100
bb	11100
ba	101
bc	11101
ca	110
cb	11110
cc	11111

Letters	Code	Letters	Code	Letters	Code
aaa	00	*baa*	1001	*caa*	1010
aab	0100	*bab*	110011	*cab*	110101
aac	0111	*bac*	110100	*cac*	11011
aba	011	*bba*	111000	*cba*	11101
abb	1011	*bbb*	1111010	*cbb*	1111101
abc	110000	*bbc*	1111011	*cbc*	11111100
aca	1000	*bca*	111001	*cca*	111100
acb	110001	*bcb*	11111000	*ccb*	11111101
acc	110010	*bcc*	11111001	*ccc*	11111111

codeword length of Huffman codes. However, we can anticipate that better approximations to the uniform distribution can be achieved by Huffman codes if the number of letters per compression further increases.

In comparison with the Huffman code, the Fano code is easier in both analysis and implementation. As can be seen from Table 7.4 and Figure 7.6, $\Pr[S_1 = 0]$ quickly converges to $1/2$, and the assumption that the compressor output $S_1, S_2, S_3, \ldots$ is independent with uniform marginal distribution can be acceptable when ν is moderately large.

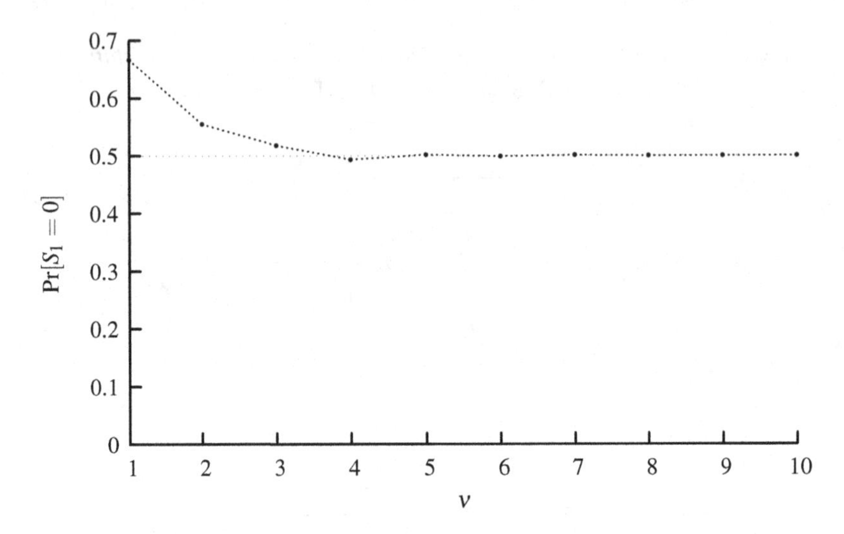

Figure 7.6 Asymptotics of ν-letter Fano codes.

7.8 Appendix: Definition of concavity

Definition 7.5 A real-valued function $h(\cdot)$ is *concave* if

$$h(\lambda p_1 + (1-\lambda)p_2) \geq \lambda h(p_1) + (1-\lambda)h(p_2) \tag{7.75}$$

for all real numbers p_1 and p_2, and all $0 \leq \lambda \leq 1$.

Geometrically, this means that the line segment that connects two points of the curve $h(\cdot)$ will always lie below the curve; see Figure 7.7 for an illustration.

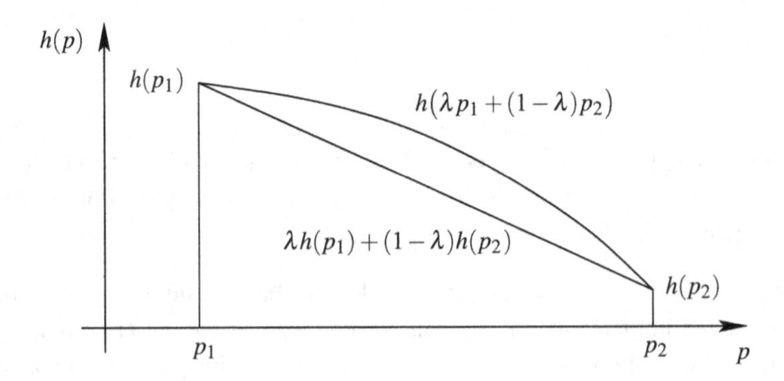

Figure 7.7 Example of a concave function.

The concavity of a function can be verified by showing that its second derivative is nonpositive. By this approach, we can prove that the binary entropy function is concave, as this can also be observed from Figure 5.2. By induction, a concave function satisfies

$$h\left(\frac{1}{k}\sum_{\ell=1}^{k} a_\ell\right) \geq \frac{1}{k}\sum_{\ell=1}^{k} h(a_\ell); \tag{7.76}$$

hence, (7.50) is also confirmed.

References

[BGT93] Claude Berrou, Alain Glavieux, and Punya Thitimajshima, "Near Shannon limit error-correcting coding and decoding: turbo-codes," in *Proceedings of IEEE International Conference on Communications (ICC)*, Geneva, Switzerland, May 23–26, 1993, pp. 1064–1070.

[Bla88] Richard E. Blahut, *Principles and Practice of Information Theory*. Addison-Wesley, Reading, MA, 1988.

[CT06] Thomas M. Cover and Joy A. Thomas, *Elements of Information Theory*, 2nd edn. John Wiley & Sons, Hoboken, NJ, 2006.

[DM98] Matthew C. Davey and David MacKay, "Low-density parity check codes over $GF(q)$," *IEEE Communications Letters*, vol. 2, no. 6, pp. 165–167, June 1998.

[For66] G. David Forney, Jr., *Concatenated Codes*. MIT Press, Cambridge, MA, 1966.

[Gal62] Robert G. Gallager, *Low Density Parity Check Codes*. MIT Press, Cambridge, MA, 1962.

[Gal68] Robert G. Gallager, *Information Theory and Reliable Communication*. John Wiley & Sons, New York, 1968.

[HW99] Chris Heegard and Stephen B. Wicker, *Turbo Coding*. Kluwer Academic Publishers, Dordrecht, 1999.

[LC04] Shu Lin and Daniel J. Costello, Jr., *Error Control Coding*, 2nd edn. Prentice Hall, Upper Saddle River, NJ, 2004.

[Sha48] Claude E. Shannon, "A mathematical theory of communication," *Bell System Technical Journal*, vol. 27, pp. 379–423 and 623–656, July and October 1948. Available: http://moser.cm.nctu.edu.tw/nctu/doc/shannon1948.pdf

[VY00] Branka Vucetic and Jinhong Yuan, *Turbo Codes: Principles and Applications*. Kluwer Academic Publishers, Dordrecht, 2000.

8
Other aspects of coding theory

We end this introduction to coding and information theory by giving two examples of how coding theory relates to quite unexpected other fields. Firstly we give a very brief introduction to the relation between Hamming codes and projective geometry. Secondly we show a very interesting application of coding to game theory.

8.1 Hamming code and projective geometry

Though not entirely correct, the concept of projective geometry was first developed by Gerard Desargues in the sixteenth century for art paintings and for architectural drawings. The actual development of this theory dated way back to the third century to Pappus of Alexandria. They were all puzzled by the axioms of Euclidean geometry given by Euclid in 300 BC who stated the following.

(1) Given any distinct two points in space, there is a unique line connecting these two points.
(2) Given any two nonparallel[1] lines in space, they intersect at a unique point.
(3) Given any two distinct parallel lines in space, they never intersect.

The confusion comes from the third statement, in particular from the concept of parallelism. How can two lines never intersect? Even to the end of universe?

[1] Note that in some Euclidean spaces we have three ways of how two lines can be arranged: they can intersect, they can be skew, or they can be parallel. For both skew and parallel lines, the lines never intersect, but in the latter situation we additionally have that they maintain a constant separation between points closest to each other on the two lines. However, the distinction between skew and parallel relies on the definition of a norm. If such a norm is not defined, "distance" is not properly defined either and, therefore, we cannot distinguish between skew and parallel. We then simply call both types to be "parallel."

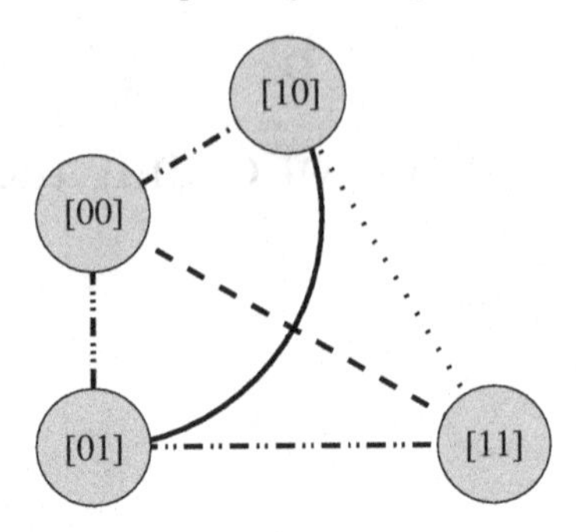

Figure 8.1 Two-dimensional binary Euclidean space.

In your daily life, the two sides of a road are parallel to each other, yet you do see them intersect at a distant point. So, this is somewhat confusing and makes people very uncomfortable. Revising the above statements gives rise to the theory of projective geometry.

Definition 8.1 (Axioms of projective geometry)

(1) Given any two distinct points in space, there is a unique line connecting these two points.
(2) Given any two distinct lines in space, these two lines intersect at a unique point.

So, all lines intersect with each other in projective geometry. For parallel lines, they will intersect at *a point at infinity*. Sounds quite logical, doesn't it? Having solved our worries, let us now focus on what the projective geometry looks like. We will be particularly working over the binary space.

Consider a two-dimensional binary Euclidean space[2] as shown in Figure 8.1. Do not worry about the fact that one line is curved and is not a straight line. We did this on purpose, and the reason will become clear later. Here we have four points, and, by the first axiom in Euclidean geometry, there can be at most $\binom{4}{2} = 6$ lines.

Exercise 8.2 *Ask yourself: why* at most *six? Can there be fewer than six lines given four points in space?* ◊

We use $[XY]$ to denote the four points in space. Consider, for example, the dash–single-dotted line $\overline{[00][10]}$ and the dash–double-dotted line $\overline{[01][11]}$. The

[2] Note that, as mentioned in Section 3.3.2, the Euclidean distance fails to work in the binary Euclidean space.

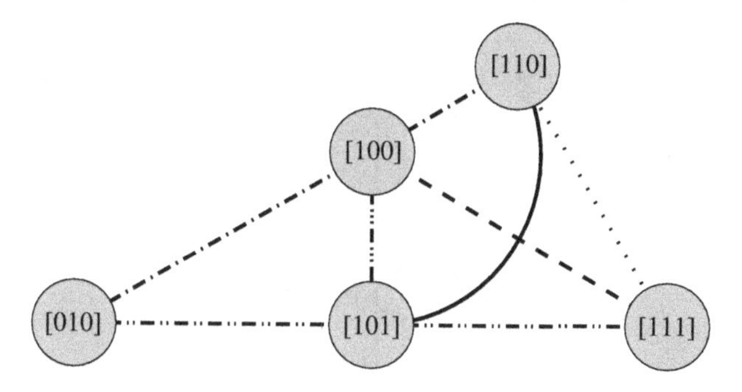

Figure 8.2 Two-dimensional binary Euclidean space with a point at infinity.

dash–single-dotted line $\overline{[00][10]}$ represents the line $Y = 0$ and the dash–double-dotted line $\overline{[01][11]}$ is the line of $Y = 1$. In Euclidean geometry, these two lines never intersect, hence it worries people. Now we introduce the concept of a "point at infinity" and make these two lines intersect as shown in Figure 8.2.

To distinguish the points at infinity from the original points, we add another coordinate Z in front of the coordinates $[XY]$. The points in the new plots are read as $[ZXY]$. The points with coordinates $[1XY]$ are the original points and the ones with $[0XY]$ are the ones at infinity. But why do we label this new point at infinity with coordinate $[010]$ and not something else? This is because points lying on the same line are *co-linear*:

$$[101] + [111] = [010], \tag{8.1}$$

i.e. we simply add the coordinates. Note that the same holds for $[100] + [110] = [010]$. Having the same result for these two sums means that the lines of $\overline{[100][110]}$ and $\overline{[101][111]}$ intersect at the same point, $[010]$.

Repeating the above process gives the geometry shown in Figure 8.3. Finally, noting that the points at infinity satisfy

$$[001] + [011] = [010], \tag{8.2}$$

we see that these three newly added points are co-linear as well. So we can add another line connecting these three points and obtain the final geometry given in Figure 8.4. This is the famous Fano plane for the two-dimensional projective binary plane. There are seven lines and seven points in this plane.

Note that the number of lines and the number of points in the projective geometry are the same, and this is no coincidence. Recall the original definition of projective geometry.

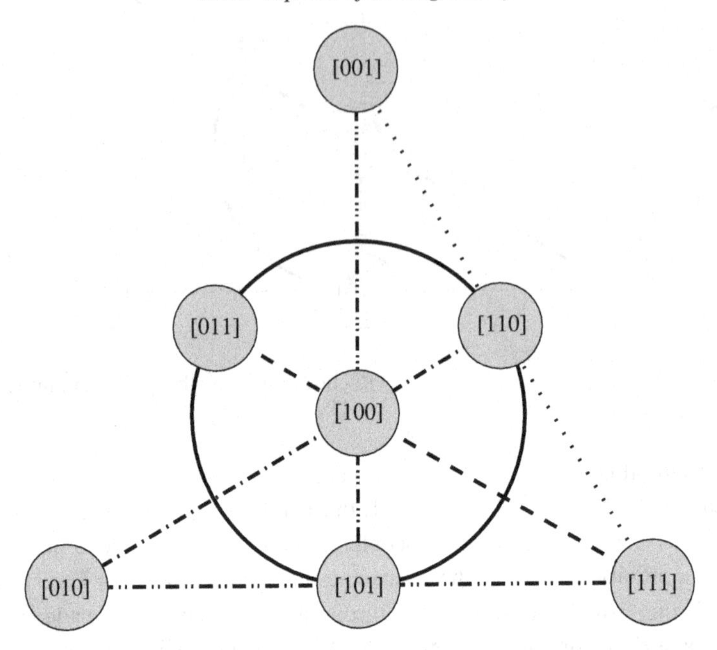

Figure 8.3 Two-dimensional binary Euclidean space with many points at infinity.

(1) Given any two distinct points in space, a unique line lies on these two points.
(2) Given any two distinct lines in space, a unique point lies on these two lines.

For the moment, forget about the literal meanings of "lines" and "points." Rewrite the above as follows.

(1) Given any two distinct □ in space, a unique ◯ lies on these two □.
(2) Given any two distinct ◯ in space, a unique □ lies on these two ◯.

So you see a symmetry between these two definitions, and this means the □s (lines) are just like the ◯s (points) and vice versa. In other words, if we label the points and lines as in Figure 8.5, we immediately discover the symmetry between the two. We only rename the L by P and the P by L in Figure 8.5(a) and Figure 8.5(b). In particular, the patterns of the lines in Figure 8.5(a) are matched to the patterns of the points in Figure 8.5(b) to signify such a duality.

To understand more about the symmetry, consider for example the following.

- Point P_1 is intersected by lines L_2, L_3, and L_5 in Figure 8.5(a). In Fig-

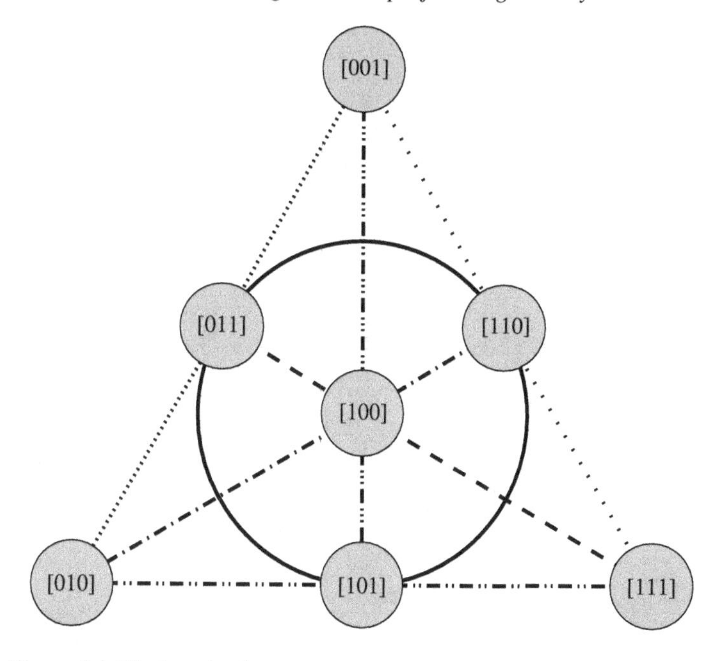

Figure 8.4 Final projective geometry of the two-dimensional binary Euclidean space with points at infinity.

ure 8.5(b), we see exactly the same relation between the lines L_3, L_5, and L_2 and the point P_1.

- Line L_1 is related to points P_2, P_3, and P_5 in Figure 8.5(b). In Figure 8.5(a), we see that L_1 passes through all these three points.

Also note that, in terms of the $[ZXY]$ axes, the lines are defined by the following functions:

$$\begin{aligned} L_1 : &\quad Z = 0, \\ L_2 : &\quad X = 0, \\ L_3 : &\quad Y = 0, \\ L_4 : &\quad Z + X = 0, \\ L_5 : &\quad X + Y = 0, \\ L_6 : &\quad Z + X + Y = 0, \\ L_7 : &\quad Z + Y = 0. \end{aligned} \tag{8.3}$$

Note that the above is quite different from what you learned in high school mathematics. For example, the function $Z = 0$ does not give a plane in projective geometry as it does in Euclidean geometry.

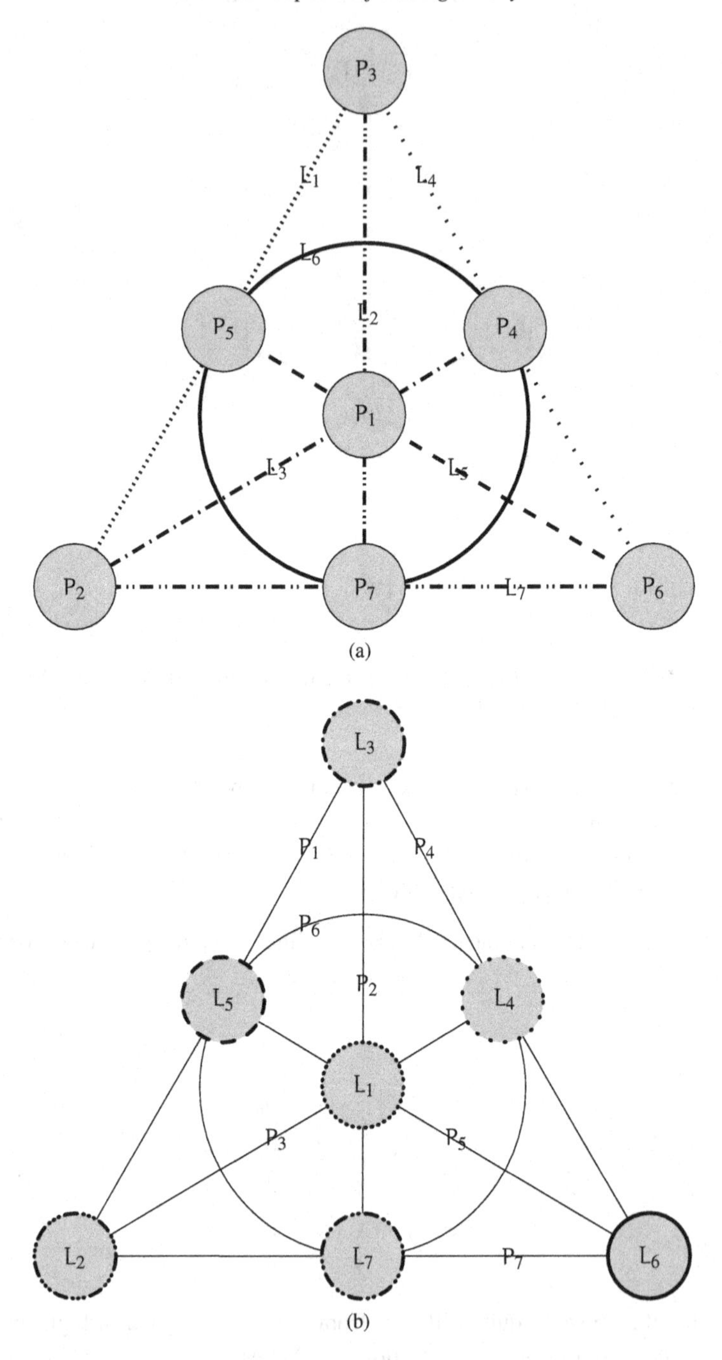

Figure 8.5 Symmetry between two definitions of projective geometry.

In general, the connection between lines in the two-dimensional Euclidean plane and the lines in the two-dimensional projective plane can be easily obtained through the following. For simplicity, let $\mathcal{E}_2$ denote the two-dimensional binary Euclidean plane, and let $\mathcal{P}_2$ denote the two-dimensional binary projective plane. Then the connection between lines in $\mathcal{E}_2$ and $\mathcal{P}_2$ is given by

$$\mathsf{L}: aX+bY+c=0 \text{ in } \mathcal{E}_2 \quad \Longleftrightarrow \quad \mathsf{L}: aX+bY+cZ=0 \text{ in } \mathcal{P}_2, \tag{8.4}$$

where not all a, b, and c equal zero.

Example 8.3 We now apply (8.4) in order to study a solid example so that we can understand more about the two geometries from the algebraic point of view. Consider, for example, lines L_3 and L_7, which represent the functions $Y=0$ and $Y=1$ in $\mathcal{E}_2$, respectively. Of course, L_3 and L_7 are parallel in $\mathcal{E}_2$ and do not intersect. On the other hand, lifting these two functions from $\mathcal{E}_2$ to $\mathcal{P}_2$ (by setting $(a,b,c)=(0\,1\,0)$ and $(0\,1\,1)$ in (8.4)) gives $Y=0$ and $Y+Z=0$, respectively. It then follows that these two functions do intersect in $\mathcal{P}_2$ at $[ZXY]=[010]$, a point satisfying these two equations. It justifies the fact that any two distinct lines always intersect with each other. An equivalent view from algebra says that every system of linear equations is always solvable in the projective sense. ◊

To relate the Fano plane to the Hamming code, we simply construct Table 8.1. A "1" means the point lies on the line, or, equivalently, that the line passes through the point.

Table 8.1 *Relation of Fano plane to Hamming code*

	P_1	P_2	P_3	P_4	P_5	P_6	P_7
L_1	0	1	1	0	1	0	0
L_2	1	0	1	0	0	0	1
L_3	1	1	0	1	0	0	0
L_4	0	0	1	1	0	1	0
L_5	1	0	0	0	1	1	0
L_6	0	0	0	1	1	0	1
L_7	0	1	0	0	0	1	1

On reading Table 8.1 row-wise and comparing these rows with codewords in Table 3.2, we see that

- line L_1 defines the codeword $(0\,1\,1\,0\,1\,0\,0)$ associated with message $(0\,1\,0\,0)$,

- line L_2 defines the codeword (1010001) associated with message (0001),
- line L_3 defines the codeword (1101000) associated with message (1000),
- line L_4 defines the codeword (0011010) associated with message (1010),
- line L_5 defines the codeword (1000110) associated with message (0110),
- line L_6 defines the codeword (0001101) associated with message (1101),
- line L_7 defines the codeword (0100011) associated with message (0011).

So, the seven lines define seven codewords of a $(7,4)$ Hamming code. What about the remaining $16-7=9$ codewords? Well, if we add an empty line, L_0, to define the codeword (00000000), then the remaining eight codewords are just the binary complement of these eight codewords. For example, the binary complement of (00000000) is (11111111), and the binary complement of (0110100) defined by L_1 is (1001011) (simply replace 0 by 1 and 1 by 0). This way you recover all the 16 codewords of the $(7,4)$ Hamming code.

While all the above seems tricky and was purposely done, it presents a means of generalization of the Hamming code. In particular, extending the two-dimensional Fano plane to higher-dimensional binary projective spaces, say $(u-1)$ dimensions,

(1) we could construct the $(2^u-1, 2^u-u-1)$ Hamming codes defined by the lines in the $(u-1)$-dimensional binary projective space,[3] and
(2) we could construct other codes defined by the s-dimensional subspaces (called s-flats or s-hyperplanes in finite geometry) in the $(u-1)$-dimensional binary projective space. These codes are therefore coined *projective geometry codes*.

Exercise 8.4 *Identify all the 16 Hamming codewords on the Fano plane plus an empty line. The points associated with each codeword form either a line or a complement of a line. Can you use this geometrical fact to decode read-outs with one-bit error? We can give you some hints for this.*

(1) Read-outs with only one nonzero position are decoded to the empty line.
(2) Read-outs with two nonzero positions are decoded to a line. Two nonzero positions mean two points in the Fano plane. The line obtained by joining these two points gives the decoded codeword. For example, if the nonzero positions are P_2 and P_4, then they form the line L_3, and the corrected output should be the codeword associated with L_3.
(3) Read-outs with three nonzero positions correspond to either a line or a triangle in the Fano plane. If it is a line, then the line gives the decoded codeword. Otherwise, the triangle can be made into a quadrangle by adding an

[3] An alternative way of getting this is given in Exercise 3.22.

extra point. Then note that the quadrangle is a complement of a line, which is a codeword. So the codeword associated with this quadrangle is the decoded output. For example, assume the nonzero positions of the read-out are P_1, P_2, *and* P_3, *which form a triangle. To make the triangle into a quadrangle, we should add point* P_6 *(note that adding either* P_4, P_5, *or* P_7 *would not work: it would still be a triangle). Then the quadrangle* $\mathsf{P}_1\mathsf{P}_2\mathsf{P}_3\mathsf{P}_6$ *is the complement of the projective line* L_6, *and hence it corresponds to a valid codeword.*

Complete the decodings of read-outs with more than three nonzero positions.

8.2 Coding and game theory

The Hamming code can be used to solve many problems in combinatorial designs as well as in game theory. One of the most famous and most interesting problems is the *hat game*. On April 10, 2001, the New York Times published an article entitled "Why mathematicians now care about their hat color." The game has the following setup.

- A team of n players enters a room, whereupon they each receive a hat with a color randomly selected from r equally probable possibilities. Each player can see everyone else's hat, but not his own.
- The players must simultaneously guess the color of their own hat, or pass.
- The team loses if any player guesses wrong or if all players pass.
- The players can meet beforehand to devise a strategy, but no communication is allowed once they are inside the room.
- The goal is to devise a strategy that gives the highest probability of winning.

Example 8.5 Let $n = 3$ and $r = 2$ with the colors Red and Blue. Let us number the three players by 1, 2, and 3, and denote their hats by H_1, H_2, and H_3, respectively. If the three players receive $(H_1, H_2, H_3) = (\text{Red}, \text{Red}, \text{Blue})$ and they guess $(\text{Red}, \text{Pass}, \text{Pass})$, then they win the game. Otherwise, for example, if they guess $(\text{Pass}, \text{Blue}, \text{Blue})$, then they lose the game due to the wrong guess of the second player. They also lose for the guess of $(\text{Pass}, \text{Pass}, \text{Pass})$. ◊

Random strategy What if the players guess at random? Say, guessing with probability $1/(r+1)$ for each color and probability $1/(r+1)$ for pass. With this random strategy, the probability of winning is given by

$$\Pr(\text{Win by using "random strategy"}) = \left(\frac{2}{r+1}\right)^n - \frac{1}{(r+1)^n}. \tag{8.5}$$

So, in Example 8.5 the random strategy will yield a probability of winning

$$\Pr(\text{Win by using "random strategy"}) = \frac{7}{27} \simeq 26\%, \tag{8.6}$$

i.e. the odds are not good.

Exercise 8.6 *Prove (8.5).*

Hint: the first term of (8.5) describes the probability of the correct color or a pass and that the second term is the probability of all passing. ◊

One-player-guess strategy Another simple strategy is to let only one of the players, say the first player, guess and let the others always pass. It is clear that if the first player passes, then the team loses. So he must make a choice. In this case, the probability of winning the game is given by

$$\Pr(\text{Win by using "one-player-guess strategy"}) = \frac{1}{r}, \tag{8.7}$$

i.e. for Example 8.5 ($r = 2$)

$$\Pr(\text{Win by using "one-player-guess strategy"}) = \frac{1}{2} = 50\%, \tag{8.8}$$

and the first player simply guesses the color to be either Red or Blue, each with probability $1/2$. This strategy is a lot better than the random guess strategy. Now the question is, can we do better than $1/2$? Actually we can, with the help of the three-times repetition code and the Hamming code we learned in Chapter 3.

Repetition code strategy For simplicity, let us focus on the case of $n = 3$ and $r = 2$ with colors being Red (denoted as binary 0) and Blue (denoted as binary 1). Recall that the three-times repetition code $\mathscr{C}_{\text{rep}}$ has two codewords $(0\,0\,0)$ and $(1\,1\,1)$. Using the repetition code, we formulate the following strategy.

- For the first player, let $(?, H_2, H_3)$ be a vector where $H_2, H_3 \in \{0, 1\}$ are the hat colors of the second and the third players, respectively. The question mark symbol "?" means that the color of the hat is unknown. The colors H_2 and H_3 are known to the first player according to the setup. Then the first player makes a guess using the following rule:

$$? = \begin{cases} 0 & \text{if } (1, H_2, H_3) \text{ is a codeword in } \mathscr{C}_{\text{rep}}, \\ 1 & \text{if } (0, H_2, H_3) \text{ is a codeword in } \mathscr{C}_{\text{rep}}, \\ \text{pass} & \text{otherwise.} \end{cases} \tag{8.9}$$

- The same applies to the second and the third players. For example, the strategy of the second player is

$$? = \begin{cases} 0 & \text{if } (H_1, 1, H_3) \text{ is a codeword in } \mathscr{C}_{\text{rep}}, \\ 1 & \text{if } (H_1, 0, H_3) \text{ is a codeword in } \mathscr{C}_{\text{rep}}, \\ \text{pass} & \text{otherwise,} \end{cases} \tag{8.10}$$

where H_1 is the color of the first player's hat known to the second player.

Example 8.7 (Continuation from Example 8.5) If the three players receive $(H_1, H_2, H_3) = (\text{Red}, \text{Red}, \text{Blue}) = (0\,0\,1)$, then

- the first player sees $(?\,0\,1)$, and neither $(0\,0\,1)$ nor $(1\,0\,1)$ is a codeword in $\mathscr{C}_{\text{rep}}$, so he passes;
- the second player sees $(0\,?\,1)$, and neither $(0\,0\,1)$ nor $(0\,1\,1)$ is a codeword in $\mathscr{C}_{\text{rep}}$, so he passes, too;
- the third player sees $(0\,0\,?)$. He notices that $(0\,0\,0)$ is a codeword in $\mathscr{C}_{\text{rep}}$, so he guesses 1.

Hence, the team wins. ◊

Exercise 8.8 *Show by listing all possibilities that the three-times repetition code strategy gives a probability of winning*

$$\Pr\big(\text{Win by using “}\mathscr{C}_{\text{rep}}\text{ code strategy”}\big) = \frac{3}{4} \tag{8.11}$$

when $n = 3$ and $r = 2$. ◊

It turns out that for $n = 3$ and $r = 2$, the three-times repetition code $\mathscr{C}_{\text{rep}}$ is the best possible strategy for this game. Also, by carrying out Exercise 8.8 you will see that the only cases for the $\mathscr{C}_{\text{rep}}$ strategy to fail are the ones when the players are given hats as $(0\,0\,0)$ and $(1\,1\,1)$, which are exactly the two codewords in $\mathscr{C}_{\text{rep}}$.

$(7,4)$ **Hamming code strategy** The $(7,4)$ Hamming code strategy is the best strategy when $n = 7$ and $r = 2$. But, prior to handing over the strategy, we quickly review what happened in the three-times repetition code case. In the previous example of $n = 3$ and $r = 2$, the ith player, given the observation $(H_1, \ldots, H_{i-1}, ?, H_{i+1}, \ldots, H_3)$, makes the following guess:

$$? = \begin{cases} 0 & \text{if } (H_1, \ldots, H_{i-1}, 1, H_{i+1}, \ldots, H_3) \text{ is a codeword in } \mathscr{C}_{\text{rep}}, \\ 1 & \text{if } (H_1, \ldots, H_{i-1}, 0, H_{i+1}, \ldots, H_3) \text{ is a codeword in } \mathscr{C}_{\text{rep}}, \\ \text{pass} & \text{otherwise.} \end{cases} \tag{8.12}$$

So, for the case of $n = 7$ and $r = 2$, let $\mathscr{C}_{\mathrm{H}}$ be the $(7,4)$ Hamming code with 16 codewords given in Table 3.2. Then we use the following similar strategy.

- The ith player, given the observation $(H_1,\ldots,H_{i-1},?,H_{i+1},\ldots,H_7)$, makes the following guess:

$$? = \begin{cases} 0 & \text{if } (H_1,\ldots,H_{i-1},1,H_{i+1},\ldots,H_7) \text{ is a codeword in } \mathscr{C}_{\mathrm{H}}, \\ 1 & \text{if } (H_1,\ldots,H_{i-1},0,H_{i+1},\ldots,H_7) \text{ is a codeword in } \mathscr{C}_{\mathrm{H}}, \\ \text{pass} & \text{otherwise.} \end{cases} \tag{8.13}$$

Example 8.9 For example, the seven players are given hats according to

$$(\text{Blue}, \text{Red}, \text{Blue}, \text{Red}, \text{Blue}, \text{Blue}, \text{Blue}) = (1\,0\,1\,0\,1\,1\,1). \tag{8.14}$$

Based on the strategy in (8.13) and the codewords of $\mathscr{C}_{\mathrm{H}}$ in Table 3.2, the players make the following guesses.

- The first player observes $(?\,0\,1\,0\,1\,1\,1)$ and notices that $(0\,0\,1\,0\,1\,1\,1)$ is a codeword. So he guesses 1, i.e. Blue.
- The second player observes $(1\,?\,1\,0\,1\,1\,1)$ and notices that neither $(1\,0\,1\,0\,1\,1\,1)$ nor $(1\,1\,1\,0\,1\,1\,1)$ are codewords. So he passes.
- You can check the remaining cases and show that they all pass.

Since the first player makes the right guess and the others pass, the team wins.

We can show the following theorem.

Theorem 8.10 *For the case of $n = 7$ and $r = 2$, the $(7,4)$ Hamming code strategy given as in (8.13) yields the probability of winning*

$$\Pr(\text{Win by using “}\mathscr{C}_{\mathrm{H}}\text{ code strategy”}) = 1 - \frac{16}{2^7} = \frac{7}{8}. \tag{8.15}$$

Proof Note that from the sphere bound of Theorems 3.20 and 3.21, we see that the $(7,4)$ Hamming code $\mathscr{C}_{\mathrm{H}}$ is a perfect packing of 16 spheres of radius 1 in the seven-dimensional binary space. Hence, given any combination of hats $\mathbf{H} = (H_1, H_2, \ldots, H_7)$, $\mathbf{H}$ must lie in one of the 16 spheres. In other words, there must exist a codeword $\mathbf{x} = (x_1,\ldots,x_7)$ of $\mathscr{C}_{\mathrm{H}}$ that is at Hamming distance at most 1 from $\mathbf{H}$. We distinguish the following cases.

Case I: If $\mathbf{H} \in \mathscr{C}_{\mathrm{H}}$, then according to the strategy (8.13), the ℓth player for every $1 \leq \ell \leq 7$ would notice that $(H_1,\ldots, H_{\ell-1},x_\ell,H_{\ell+1},\ldots,H_7)$ is a codeword in $\mathscr{C}_{\mathrm{H}}$, hence he will guess $\bar{x}_\ell$, the binary complement of x_ℓ. The team always loses in this case.

Case II: If $\mathbf{H} \notin \mathscr{C}_{\mathrm{H}}$, then $\mathbf{H}$ is at Hamming distance 1 from some codeword $\mathbf{x}$. Say the difference is at the jth place, for some j, i.e. the hat color H_j of the jth player equals $\bar{x}_j$.

- For the ℓth player, $\ell \neq j$, we see from strategy (8.13) that $(H_1, \ldots, H_{\ell-1}, x_\ell, H_{\ell+1}, \ldots, H_7)$ is at Hamming distance 1 from $\mathbf{x}$ and $(H_1, \ldots, H_{\ell-1}, \bar{x}_\ell, H_{\ell+1}, \ldots, H_7)$ is at Hamming distance 2 from $\mathbf{x}$. Both cannot be codewords because the codewords of Hamming code are separated by a distance of at least 3. Thus, the ℓth player always passes in this case.
- The jth player observes that $(H_1, \ldots, H_{j-1}, x_j, H_{j+1}, \ldots, H_7) = \mathbf{x}$ is a codeword, hence he guesses $\bar{x}_j$, which is the correct guess.

Thus, the team wins.

From the above analysis we see that the team loses if, and only if, $\mathbf{H}$ is a codeword in $\mathscr{C}_{\mathrm{H}}$. Since there are 16 such possibilities, we conclude that

$$\Pr(\text{Lose by using "}\mathscr{C}_{\mathrm{H}}\text{ code strategy"}) = \frac{16}{2^7} \tag{8.16}$$

and the theorem is proven. □

The strategy we have devised above is related to the *covering* of error-correcting codes. The concept of covering is the opposite of that of sphere packing: the problem of covering asks what the minimum number of t is such that the radius-t spheres centered at the 2^k codewords of a code fill up the complete n-dimensional binary space. Here the spheres are allowed to overlap with each other. The three-times repetition code $\mathscr{C}_{\mathrm{rep}}$ and the Hamming code $\mathscr{C}_{\mathrm{H}}$ are both 1-covering codes because radius-1 spheres centered at their codewords completely cover the three-dimensional and seven-dimensional binary space, respectively.

In general, we can show the following theorem.

Theorem 8.11 *Let $\mathscr{C}$ be a length-n 1-covering error-correcting code with $|\mathscr{C}|$ codewords. Then for the hat game with n players and $r = 2$ colors, following the strategy defined by $\mathscr{C}$ as in (8.13), yields a winning probability of*

$$\Pr(\text{Win by using "}\mathscr{C}\text{ code strategy"}) = 1 - \frac{|\mathscr{C}|}{2^n}. \tag{8.17}$$

Exercise 8.12 *Prove Theorem 8.11 by showing that (Case I) if $\mathbf{H} \in \mathscr{C}$, the team always loses, and (Case II) if $\mathbf{H} \notin \mathscr{C}$, the team always wins even if the codewords of the 1-covering code are not separated by a distance of at least 3.*

Hint: $(H_1, \ldots, H_{\ell-1}, \bar{x}_\ell, H_{\ell+1}, \ldots, H_n)$ could be a codeword. ◇

Finally we remark that both $\mathscr{C}_{\text{rep}}$ and $\mathscr{C}_{\text{H}}$ are optimal 1-covering codes because they have the smallest possible code size among all 1-covering codes of length 3 and length 7, respectively. The fact that the three-times repetition code $\mathscr{C}_{\text{rep}}$ is an optimal length-3 1-covering code follows from the third case in Theorem 3.21 with $u = 1$.

8.3 Further reading

In this chapter we have briefly discussed two different aspects of coding theory. Using the $(7,4)$ Hamming code as a starting example, we have shown how the error-correcting codes can be used in the study of finite geometry as well as game theory. To encourage further investigations in this direction, we provide a short list of other research fields that are closely related to coding theory.

Cryptography One aim in cryptography is message encryption so that eavesdroppers cannot learn the true meaning of an encrypted message. The encryption device has a key, which is known to the sender and the recipient, but not to the eavesdropper. Given the key $\mathbf{K}$, the encryption device encrypts plaintext $\mathbf{S}$ into ciphertext $\mathbf{C}$. It is hoped that without the key the eavesdropper cannot easily recover the plaintext $\mathbf{S}$ from $\mathbf{C}$. In 1949 Shannon [Sha49] first applied information theory to the study of cryptography and defined the notion of *perfect secrecy*. We say that the communication is perfectly secure if the mutual information between $\mathbf{S}$ and $\mathbf{C}$ is zero, i.e. $I(\mathbf{S};\mathbf{C}) = 0$. Noting that

$$I(\mathbf{S};\mathbf{C}) = H(\mathbf{S}) - H(\mathbf{S}|\mathbf{C}), \tag{8.18}$$

a perfectly secure communication means that the eavesdropper can never learn *any* information about $\mathbf{S}$ from the observation of $\mathbf{C}$. While none of the currently used cryptosystems can offer such perfect secrecy, in 1978 Robert J. McEliece proposed a highly secure cryptosystem based on the use of (n,k) binary linear error-correcting codes. McEliece's cryptosystem with large n is immune to all known attacks, including those made by quantum computers. Readers interested in this line of research are referred to [McE78] and [Sti05] for further reading.

Design of pseudo-random sequences The pseudo-random number generator is perhaps one of the most important devices in modern computing. A possible implementation of such a device is through the use of *maximum-length sequences*, also known as *m-sequences*. The m-sequence is a binary pseudo-random sequence in which the binary values 0 and 1 appear almost statistically independent, each with probability $1/2$. Given the initial seed, the m-sequence can be easily generated by feedback shift registers. It is also one of the key components in modern cellular communication systems that are built upon code-division multiple-access (CDMA) technology. The m-sequence and the Hamming code are closely connected. In fact, the m-sequence is always a codeword in the dual of the Hamming code. Readers are referred to [McE87] and [Gol82] for more details about this connection and about the design of pseudo-random sequences.

Latin square and Sudoku puzzle The Latin square is a special kind of combinatorial object which many people have seen in some mathematical puzzles. Specifically, a Latin square is an $(n \times n)$ array in which each row and each column consist of the same set of elements without repetition. For example, the following is a (3×3) Latin square.

$$\begin{bmatrix} 3 & 1 & 2 \\ 1 & 2 & 3 \\ 2 & 3 & 1 \end{bmatrix} \tag{8.19}$$

The famous game of Sudoku can also be regarded as a special kind of (9×9) Latin square. Sudoku puzzles are probably the most popular among all Latin squares. Another interesting extension is called the *orthogonal array*, which has very useful applications in software testing. Two $(n \times n)$ Latin squares A and B are said to be orthogonal if all the n^2 pairs $([\mathsf{A}]_{i,j}, [\mathsf{B}]_{i,j})$ are distinct. For example, the following (4×4) Latin squares are orthogonal to each other:

$$\begin{bmatrix} 1 & 2 & 3 & 4 \\ 2 & 1 & 4 & 3 \\ 3 & 4 & 1 & 2 \\ 4 & 3 & 2 & 1 \end{bmatrix} \quad \text{and} \quad \begin{bmatrix} 3 & 4 & 1 & 2 \\ 4 & 3 & 2 & 1 \\ 2 & 1 & 4 & 3 \\ 1 & 2 & 3 & 4 \end{bmatrix}. \tag{8.20}$$

While there are many ways to construct mutually orthogonal arrays, one of the most notable constructions is from the finite projective plane we studied in Section 8.1. A famous theorem in this area states that there exists $(n-1)$ mutually orthogonal $(n \times n)$ Latin squares if, and only if, there exists a finite projective plane in which every projective line has $(n-1)$ points. Again, the finite projective planes are tied closely to the Hamming codes. Please refer to [Bry92] and [vLW01] for a deeper discussion.

Balanced incomplete block designs The problem with block design is as follows: v players form t teams with m members in each team. Two conditions are required: (a) each player is in precisely μ teams, and (b) every pair of players is in precisely λ teams. Configurations meeting the above requirements are termed (v,t,μ,m,λ) *block designs*. It should be noted that these parameters are not all independent. The main challenge is, given a set of parameters, to find out whether the design exists, and, if the answer is yes, how to construct it. For many parameters these questions are still unanswered. The (v,t,μ,m,λ) block designs have many applications to experimental designs, cryptography, and optical fiber communications. Moreover, block designs can be transformed into a class of error-correcting codes, termed *constant-weight codes*. Certain block-designs with $\lambda = 1$ can be obtained from finite projective planes. For more details please refer to [HP03].

References

[Bry92] Victor Bryant, *Aspects of Combinatorics: A Wide-Ranging Introduction*. Cambridge University Press, Cambridge, 1992.

[Gol82] Solomon W. Golomb, *Shift Register Sequences*, 2nd edn. Aegean Park Press, Laguna Hills, CA, 1982.

[HP03] W. Cary Huffman and Vera Pless, eds., *Fundamentals of Error-Correcting Codes*. Cambridge University Press, Cambridge, 2003.

[McE78] Robert J. McEliece, "A public-key cryptosystem based on algebraic coding theory," DSN Progress Report 42-44, Technical Report, January and February 1978.

[McE87] Robert J. McEliece, *Finite Field for Scientists and Engineers*, Kluwer International Series in Engineering and Computer Science. Kluwer Academic Publishers, Norwell, MA, 1987.

[Sha49] Claude E. Shannon, "Communication theory of secrecy systems," *Bell System Technical Journal*, vol. 28, no. 4, pp. 656–715, October 1949.

[Sti05] Douglas R. Stinson, *Cryptography: Theory and Practice*, 3rd edn. Chapman & Hall/CRC Press, Boca Raton, FL, 2005.

[vLW01] Jacobus H. van Lint and Richard M. Wilson, *A Course in Combinatorics*, 2nd edn. Cambridge University Press, Cambridge, 2001.

References

[BGT93] Claude Berrou, Alain Glavieux, and Punya Thitimajshima, "Near Shannon limit error-correcting coding and decoding: turbo-codes," in *Proceedings of IEEE International Conference on Communications (ICC)*, Geneva, Switzerland, May 23–26, 1993, pp. 1064–1070.

[Bla88] Richard E. Blahut, *Principles and Practice of Information Theory*. Addison-Wesley, Reading, MA, 1988.

[Bry92] Victor Bryant, *Aspects of Combinatorics: A Wide-Ranging Introduction*. Cambridge University Press, Cambridge, 1992.

[BT02] Dimitri P. Bertsekas and John N. Tsitsiklis, *Introduction to Probability*. Athena Scientific, Belmont, MA, 2002.

[CA05] Po-Ning Chen and Fady Alajaji, *Lecture Notes on Information Theory*, vol. 1, Department of Electrical Engineering, National Chiao Tung University, Hsinchu, Taiwan, and Department of Mathematics & Statistics, Queen's University, Kingston, Canada, August 2005. Available: http://shannon.cm.nctu.edu.tw/it/itvol12004.pdf

[CS99] John Conway and Neil J. A. Sloane, *Sphere Packings, Lattices and Groups*, 3rd edn. Springer Verlag, New York, 1999.

[CT06] Thomas M. Cover and Joy A. Thomas, *Elements of Information Theory*, 2nd edn. John Wiley & Sons, Hoboken, NJ, 2006.

[DM98] Matthew C. Davey and David MacKay, "Low-density parity check codes over $GF(q)$," *IEEE Communications Letters*, vol. 2, no. 6, pp. 165–167, June 1998.

[Fan49] Robert M. Fano, "The transmission of information," Research Laboratory of Electronics, Massachusetts Institute of Technology (MIT), Technical Report No. 65, March 17, 1949.

[For66] G. David Forney, Jr., *Concatenated Codes*. MIT Press, Cambridge, MA, 1966.

[Gal62] Robert G. Gallager, *Low Density Parity Check Codes*. MIT Press, Cambridge, MA, 1962.

[Gal68] Robert G. Gallager, *Information Theory and Reliable Communication*. John Wiley & Sons, New York, 1968.

[Gal01] Robert G. Gallager, "Claude E. Shannon: a retrospective on his life, work, and impact," *IEEE Transactions on Information Theory*, vol. 47, no. 7, pp. 2681–2695, November 2001.

[Gol49] Marcel J. E. Golay, "Notes on digital coding," *Proceedings of the IRE*, vol. 37, p. 657, June 1949.

[Gol82] Solomon W. Golomb, *Shift Register Sequences*, 2nd edn. Aegean Park Press, Laguna Hills, CA, 1982.

[Har28] Ralph Hartley, "Transmission of information," *Bell System Technical Journal*, vol. 7, no. 3, pp. 535–563, July 1928.

[HP98] W. Cary Huffman and Vera Pless, eds., *Handbook of Coding Theory*. North-Holland, Amsterdam, 1998.

[HP03] W. Cary Huffman and Vera Pless, eds., *Fundamentals of Error-Correcting Codes*. Cambridge University Press, Cambridge, 2003.

[HW99] Chris Heegard and Stephen B. Wicker, *Turbo Coding*. Kluwer Academic Publishers, Dordrecht, 1999.

[Kar39] William Karush, "Minima of functions of several variables with inequalities as side constraints," Master's thesis, Department of Mathematics, University of Chicago, Chicago, IL, 1939.

[Khi56] Aleksandr Y. Khinchin, "On the fundamental theorems of information theory," (in Russian), *Uspekhi Matematicheskikh Nauk XI*, vol. 1, pp. 17–75, 1956.

[Khi57] Aleksandr Y. Khinchin, *Mathematical Foundations of Information Theory*. Dover Publications, New York, 1957.

[KT51] Harold W. Kuhn and Albert W. Tucker, "Nonlinear programming," in *Proceedings of Second Berkeley Symposium on Mathematical Statistics and Probability*, J. Neyman, ed. University of California Press, Berkeley, CA, 1951, pp. 481–492.

[LC04] Shu Lin and Daniel J. Costello, Jr., *Error Control Coding*, 2nd edn. Prentice Hall, Upper Saddle River, NJ, 2004.

[Mas96] James L. Massey, *Applied Digital Information Theory I and II,* Lecture notes, Signal and Information Processing Laboratory, ETH Zurich, 1995/1996. Available: http://www.isiweb.ee.ethz.ch/archive/massey_scr/

[McE78] Robert J. McEliece, "A public-key cryptosystem based on algebraic coding theory," DSN Progress Report 42-44, Technical Report, January and February 1978.

[McE85] Robert J. McEliece, "The reliability of computer memories," *Scientific American*, vol. 252, no. 1, pp. 68–73, 1985.

[McE87] Robert J. McEliece, *Finite Field for Scientists and Engineers*, Kluwer International Series in Engineering and Computer Science. Kluwer Academic Publishers, Norwell, MA, 1987.

[MS77] F. Jessy MacWilliams and Neil J. A. Sloane, *The Theory of Error-Correcting Codes*. North-Holland, Amsterdam, 1977.

[Nor89] Arthur L. Norberg, "An interview with Robert M. Fano," Charles Babbage Institute, Center for the History of Information Processing, April 1989.

[Pin54] Mark S. Pinsker, "Mutual information between a pair of stationary Gaussian random processes," (in Russian), *Doklady Akademii Nauk SSSR*, vol. 99, no. 2, pp. 213–216, 1954, also known as "The quantity of information about a Gaussian random stationary process, contained in a second process connected with it in a stationary manner."

[Ple68] Vera Pless, "On the uniqueness of the Golay codes," *Journal on Combination Theory*, vol. 5, pp. 215–228, 1968.

[RG04] Mohammad Rezaeian and Alex Grant, "Computation of total capacity for discrete memoryless multiple-access channels," *IEEE Transactions on Information Theory*, vol. 50, no. 11, pp. 2779–2784, November 2004.

[Say99] Jossy Sayir, "On coding by probability transformation," Ph.D. dissertation, ETH Zurich, 1999, Diss. ETH No. 13099. Available: http://e-collection.ethbib.ethz.ch/view/eth:23000

[Sha37] Claude E. Shannon, "A symbolic analysis of relay and switching circuits," Master's thesis, Massachusetts Institute of Technology (MIT), August 1937.

[Sha48] Claude E. Shannon, "A mathematical theory of communication," *Bell System Technical Journal*, vol. 27, pp. 379–423 and 623–656, July and October 1948. Available: http://moser.cm.nctu.edu.tw/nctu/doc/shannon1948.pdf

[Sha49] Claude E. Shannon, "Communication theory of secrecy systems," *Bell System Technical Journal*, vol. 28, no. 4, pp. 656–715, October 1949.

[Sha56] Claude E. Shannon, "The zero error capacity of a noisy channel," *IRE Transactions on Information Theory*, vol. 2, no. 3, pp. 8–19, September 1956.

[Sti91] Gary Stix, "Profile: Information theorist David A. Huffman," *Scientific American (Special Issue on Communications, Computers, and Networks)*, vol. 265, no. 3, September 1991.

[Sti05] Douglas R. Stinson, *Cryptography: Theory and Practice*, 3rd edn. Chapman & Hall/CRC Press, Boca Raton, FL, 2005.

[Tie73] Aimo Tietäväinen, "On the nonexistence of perfect codes over finite fields," *SIAM Journal on Applied Mathematics*, vol. 24, no. 1, pp. 88–96, January 1973.

[Tun67] Brian P. Tunstall, "Synthesis of noiseless compression codes," Ph.D. dissertation, Georgia Institute of Technology, September 1967.

[vLW01] Jacobus H. van Lint and Richard M. Wilson, *A Course in Combinatorics*, 2nd edn. Cambridge University Press, Cambridge, 2001.

[VY00] Branka Vucetic and Jinhong Yuan, *Turbo Codes: Principles and Applications*. Kluwer Academic Publishers, Dordrecht, 2000.

[WD52] Philip M. Woodward and Ian L. Davies, "Information theory and inverse probability in telecommunication," *Proceedings of the IEE*, vol. 99, no. 58, pp. 37–44, March 1952.

[Wic94] Stephen B. Wicker, *Error Control Systems for Digital Communication and Storage*. Prentice Hall, Englewood Cliffs, NJ, 1994.

Index

Italic entries are to names.

Printed in the United States
By Bookmasters